Ecology, Recreation and Tourism

Published by the Press Syndicate of the University of Cambridge
The Pitt Building, Trumpington Street, Cambridge CB2 1RP
32 East 57th Street, New York, NY 10022, USA
10 Stamford Road, Oakleigh, Melbourne 3166, Australia

First published 1986

Printed in Great Britain by the University Press, Cambridge

British Library cataloguing in publication data
Edington, John M.
Ecology, recreation and tourism.
1. Tourist trade—Environmental aspects
I. Title   II. Edington, M. Ann
338.4′791   G156.5.E5

Library of Congress cataloguing in publication data
Edington, John M.
Ecology, recreation, and tourism.
Bibliography: p. 181
Includes index.
1. Outdoor recreation—Environmental aspects.
2.Tourist trade—Environmental aspects. 3. Nature
conservation.   I. Edington, M. Ann.   II. Title.
QH545.087E35   1986   380.1′459104   85-30944

ISBN 0 521 30646 9 hard covers
ISBN 0 521 31409 7 paperback

UP

CONTENTS

Acknowledgements   ·ii

1. Introduction   1
2. Active physical pursuits   13
3. Observing wildlife   34
4. Recreational hunting and fishing   51
5. Enjoyment of scenery   77
6. Disease hazards   96
7. Insect nuisances   118
8. Hazards associated with larger animals   135
9. Environmental effects of tourist support facilities   167

References   181
Index   198

ACKNOWLEDGEMENTS

Our interest in the ecological aspects of tourism, and the realisation that there could be negative as well as positive implications, was first stimulated by visits to National Parks in East Africa in 1978. Since that time we have had the opportunity to explore these issues in a number of other countries.

In North America we were able to use the University of California, Berkeley, as a base, and we are especially grateful to Professor Joe McBride of the Department of Forestry and Resource Management for his advice and hospitality. Our exploration of the disease hazards faced by tourists has been greatly assisted by visits to the Faculty of Veterinary Medicine & Animal Science at the Agricultural University of Malaysia (Universiti Pertanian Malaysia) and the Institute of Medical Research in Kuala Lumpur. These visits were partly funded by the Inter-University Council for Higher Education Overseas, and the British Council, for whose support we are grateful. We were able to form a clearer picture of potential hazards posed by marine animals, and the problems of tourist developments in coastal regions, as a result of useful discussions with the Queensland Fisheries Department, and the National Parks and Wildlife Service in Brisbane, Australia. In Britain, we have been greatly assisted in exploring some specific instances of interactions between recreational and conservation interests, by having access to information collected by the Nature Conservancy Council.

In preparing the book for publication we owe a special debt to

Marion Eynon for her patience and care in producing the typescript. A number of individuals and organisations have generously provided photographic illustrations. These contributions are detailed in the photographic captions. Photographs appearing without specific acknowledgement are the work of M.A.E. Virtually all the line diagrams in the book have been drawn specifically for this purpose by J.M.E. and he is grateful to Pam Davies for suggestions about the layout of some of these. We also wish to thank the following for permission to reproduce existing diagrams: the United Nations Environment Programme for Fig. 1.1; Dr R. T. Paine and the American Association for the Advancement of Science for Fig. 1.4 from *Science*; Dr T. Weaver and Blackwell Scientific Publications for Fig. 5.1 from the *Journal of Applied Ecology*; and Dr H. J. Mader and Elsevier Applied Science Publishers for Fig. 9.1 from *Biological Conservation*.

Finally we would like to acknowledge the facilities and assistance afforded to us at University College, Cardiff.

J. M. Edington
M. A. Edington
October 1985

1.
INTRODUCTION

Recreation and its travel-based variant, tourism, undoubtedly serve an important social function by providing a contrast with everyday preoccupations. Moreover there are several factors which make it likely that the present expansion of recreational activities will continue. These factors include longer holiday entitlements, larger disposable incomes and a greater expectation of life after retirement. One can also predict confidently that technical advances in the fields of air transport and telecommunications will further extend the scope of international tourism.

Recognising the benefits of recreation and tourist development should not, however, disguise the fact that such enterprises can also produce undesirable side effects. A United Nations study of the impact of tourism on developing countries has pointed out that, whilst tourism frequently confers economic advantages on the host country and promotes cultural contacts between different societies, it can also be the cause of environmental and social disruption (Fig. 1.1).

Finding ways to maximise the benefits of recreation and tourism whilst at the same time minimising their adverse side-effects necessarily involves contributions from a wide range of disciplines, including economics, sociology, architecture, engineering, geography and biology. The nature of most of these inputs has already been examined in a number of general texts. Thus Baud-Bovy & Lawson (1977) have reviewed the structural and architectural aspects, and Burkart & Medlik (1974) the economic and organisational approach. The

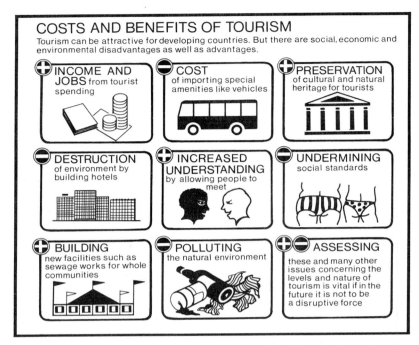

Fig. 1.1. Diagram illustrating possible costs and benefits of tourism (reproduced with permission from '*Costs and Benefits of Tourism*', United Nations Environment Programme 1979).

interactions between socio–economic and geographical factors has been explored by a series of authors including Simmons (1975), Coppock & Duffield (1975), Pearce (1981), Mathieson & Wall (1982), Smith (1983) and Patmore (1983). The particular problems raised by the interactions between tourists and host communities have been discussed by Pearce (1982), and in the collections of papers edited by Smith (1978) and De Kadt (1979).

Compared with the treatment of these topics, little comprehensive attention has been given to the contribution of the biological sciences. This is a serious omission. An examination of the scattered literature makes it clear that a proper understanding of biological, or more specifically, ecological factors can significantly reduce the scale of environmental damage associated with recreational and tourist development. Equally important it can serve to dissuade developers

2

from embarking on schemes whose economic viability is likely to be jeopardised by adverse ecological influences.

Recreational damage

Taking first the question of environmental damage, one can readily imagine that many of the newer mechanical devices used in recreation, such as trail bikes and snowmobiles would represent a source of environmental disturbance. Ecological studies confirm this prediction and draw attention to the danger of underestimating the scale of the damage, much of which occurs below the surface and out of sight. Participants in pursuits such as animal-watching might reasonably be expected to have a special regard for the welfare of their animal subjects. Unfortunately this is not invariably the case. Both competition between participants and a failure to perceive the fragility of behavioural and ecological relationships, often lead to serious disturbances of wildlife populations. Similar problems arise in relation to recreational hunting and fishing. Although the modern hunter habitually safeguards the interests of his chosen quarry species, his single-minded attention to this aim frequently produces environmental disruption in other directions. The introduction of alien species into new habitats, the unnecessary persecution of predatory species and problems caused by spent lead shot and discarded fishing weights are all examples of the wider ramifications of these activities. Finally even the ostensibly passive occupation of enjoying beautiful scenery can have self-defeating consequences when it results in unsightly damage to vegetation and soils in scenic areas. These various issues form the subject matter of Chapters 2–5.

Constraints on recreation and tourism imposed by biological agents

The second category of biological inputs relates to the contrasting situation in which man is the potential recipient and other organisms are the generators of adverse affects. As the growth of tourism takes holidaymakers into increasingly remote and challenging environments disease problems are being encountered which were previously the concern only of occupational groups such as foresters and prospectors.

Problems also arise from the increasing popularity of holidays in the tropics which leads to visitors from temperate countries becoming involved in disease cycles based on endemic infections in local human populations. Other difficulties generated by biological agents concern the nuisances caused by swarming and biting insects, and the potential hazards associated with larger animals such as sharks, bears and snakes. These various issues are explored in Chapters 6, 7 and 8.

The design of tourist resorts and associated support facilities

Whilst architectural and engineering considerations must be expected to have a decisive influence on the design of tourist resorts, failure to take account of ecological factors can have serious consequences. Striking examples are provided by situations where poorly designed organic-waste disposal facilities have allowed the proliferation of troublesome plant and animal species. These have variously interfered with water-based sports such as swimming and angling, and have generated a wide variety of nuisances and health hazards. It is therefore in the recreational planner's own interest to take these factors fully into account.

It is also important to try to ensure, particularly at resorts established in previously remote areas, that service provisions for vehicular traffic, waste disposal, water and energy supplies are designed to avoid disruption of natural systems. Apart from their intrinsic value, these systems frequently contribute significantly to the attractions of tourist sites. These issues are examined in Chapter 9.

The scope and application of ecology

In the context of planning for recreation and tourism, the most relevant aspect of biology is that known as ecology. Ecology concentrates on the environmental relationships of organisms, their reactions with one another and the properties of whole assemblages of organisms (communities) occupying the same habitat. It is useful for our present purpose to review briefly the principal concepts arising from the work of ecologists.

Relationships with physical factors

Species are frequently restricted in their distribution by physical factors in the environment. This phenomenon is well illustrated by the limitation of aquatic animals to particular ranges of salinity, water flow rate or temperature. Many animals also show distinct preferences for particular plant formations. Apart from any significance as food sources, plants frequently represent part of the physical architecture of the environment and serve to provide shelter and support. As will become apparent, defining the precise physical requirements of organisms can be useful both in predicting the disruptive effects of recreational activities, and also in identifying the situations likely to favour troublesome or dangerous species.

Organism-to-organism relationships

Relationships between different organisms are of three main types (Fig. 1.2). Firstly there are interactions between members of the same species (a). Under this heading ecologists are concerned with such matters as territorial organisation, breeding behaviour and maternal care. Secondly there are interactions between species whose needs are sufficiently similar for them to be in potential competition (b). The main interest here lies in trying to identify the mechanisms operating to reduce the incidence of serious strife. Usually this is a matter of the potential protagonists subdividing the resources of the habitat in some way. Thirdly there are the so called 'food chain' relationships in which one species has a direct nutritive relationship with another. The relationship between a plant-feeding animal (herbivore) and its plant food, or a predator and its prey are familiar examples in this category (c, d).

If one turns to parasitic organisms more complicated nutritive relationships frequently become apparent. For example when a blood-feeding insect such as a mosquito feeds on man, it is not only acting as a parasite itself, it may simultaneously transmit another parasite, such as the virus which causes yellow fever. To complicate the situation further, wild species in the area, for example monkeys, may also be bitten and act as *disease reservoirs* (Fig. 1.3). The disease-transmitting species, in this example is known as a *vector*.

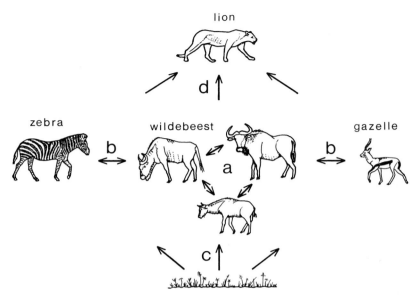

Fig. *1.2*. Diagram illustrating the various relationships between organisms: (a) interactions between members of the same species; (b) interactions between different species which are potential competitors; (c) herbivore/plant relationships; (d) predator/prey relationships (example based on an East African grassland community investigated by Bell 1971).

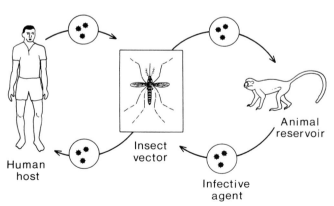

Fig. *1.3*. Diagram showing the range of inter-species relationships which can be involved in a disease transmission chain (example based on the transmission of yellow fever in rural parts of Africa).

Studies in population dynamics have the objective of analysing organism-to-organism relationships in quantitative and functional terms (Moss *et al.* 1982). They seek especially to reveal how the population size of a given species is affected by the influence of factors such as competition, predation and food supply, as well as social interactions within the species itself.

It will be seen that studies of organism-to-organism relationships have a wide application to recreational situations. The proper regulation of recreational hunting requires a basic understanding of the principles of population dynamics. The perpetuation of the open grassy habitats widely used for recreation in temperate regions requires an appreciation of the particular herbivore / plant relationship represented by the interaction between grazing stock and the vegetation on which they feed. Reducing the problem of disturbances caused by animal-watchers needs an awareness of the frequently delicate balances between predator and prey species, and the subtle interactions between newly born animals and their parents.

Community structure

Encouraged by the progress made in understanding inter-organism relationships, ecologists have been increasingly turning their attention to the functioning of whole communities (Giller 1984), a community being defined for this purpose as the total assemblage of species occupying a particular habitat. The wider term *ecosystem* is also used in this context to cover not only the biological populations present at a site but also the totality of chemical and physical influences operating there.

Species living in long-established communities are frequently revealed as being in a condition of dynamic balance with one another. Prey species commonly develop strategies to prevent their populations from being decimated by predators. Likewise organisms with broadly similar needs typically reduce the intensity of competition by evolving subtly different ways of using the environment. Alternately, potentially competing species are sometimes prevented from serious conflict because the action of predators checks the growth of their populations. This is the scientific basis for the popular notion of a 'balance of nature'.

7

In these circumstances it is not surprising that the entry of an alien species into a long-established and stable community can have seriously disturbing consequences. Striking examples of this kind of disruption are provided by the effects of introducing domesticated species to islands (Coblentz 1978). Goats for example frequently cause extensive damage to native vegetation, much of which is ill-adapted to resist their attacks, and at the same time reduce the availability of plant food to valuable indigenous herbivores such as the giant tortoises on the Galápagos Islands.

A dramatic illustration of the effects of an alien species on an aquatic habitat is provided by the introduction of the 'peacock bass' *Cichla ocellaris*) into Gatun Lake in Central America (Zaret & Paine 1973). This fish is a native of the River Amazon but was taken to Central America by fish breeders and anglers. Its presence has completely transformed the existing lake community (Fig. 1.4). Previously-abundant fish species have now disappeared as a result of predation and competition from *Cichla*, and in turn this has had the effect of depriving larger predatory fish and birds of a food source.

This risk of community disruption by introduced species has a direct bearing on the conduct of recreational hunting and fishing enterprises. As will be seen, there are now many documented case histories which demonstrate the serious consequences of introducing species into new geographical areas for sporting purposes.

The removal of a species from a balanced community can have equally far-reaching effects, particularly if the species in question had previously been playing a key role in the operation of the system (Paine 1966, 1980). For example when limpets (*Patella vulgata*) are removed from a seashore, either intentionally as part of an experiment, or accidentally in the aftermath of an oil-spill, the loss of their feeding activity can cause previously inconspicuous seaweed populations to increase dramatically and monopolise the beach (Jones 1948, Southward & Southward 1978).

A particularly striking example of the side effects of species removal is provided by the extermination of the sea otter (*Enhydra lutris*) from parts of its original range by hunters and fur traders. From historical reconstructions and studies on modern marine communities (Estes & Palmisano 1974, Estes *et al.* 1978, Simenstad *et al.* 1978) it is possible to infer that the removal of the otters allowed sea urchins, which are

a

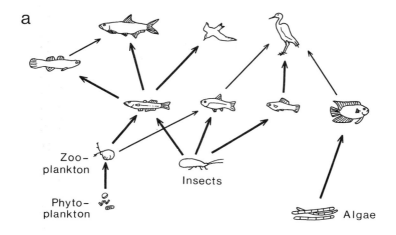

Zoo-
plankton

Insects

Phyto-
plankton

Algae

b

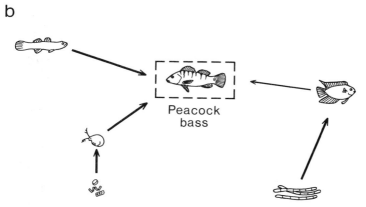

Peacock
bass

Fig. 1.4. The disruptive effects of introducing an alien species, the 'peacock bass' (*Cichla ocellaris*) into a Central American lake community (food-chain species not to scale) (adapted with permission from 'Species introduction in a tropical lake', T. M. Zaret & T. T. Paine, *Science* **182**, 2 November 1973, pp. 449–55. Copyright by the American Association for the Advancement of Science).

one of their favoured foods, to increase dramatically. These dense populations of sea urchins then proceeded to feed voraciously on the giant seaweed or 'kelp' beds in the area, and virtually to destroy them (Fig. 1.5b). This in turn adversely affected the fish species which used the seaweed beds for shelter, or were dependent on the food chain

9

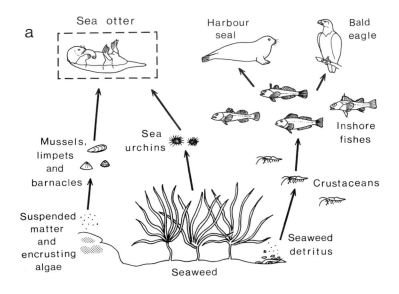

a

b

Fig. 1.5. The consequences of removing a key species, the sea otter (*Enhydra lutris*) from a marine community in the North Pacific (food chain species not to scale) (based on studies by Estes *et al.* 1974, 1978, Simenstad *et al.* 1978).

generated by the breakdown of seaweed detritus. The final effect in the sequence was for fish-eating species such as sea eagles and seals to be deprived of their normal prey.

Cautionary tales of this kind need to be born in mind whenever there are suggestions that supposedly-dangerous species should be removed from a habitat in the interests of public safety. If the species in question happens to be playing a key role in the functioning of the community, there is the risk that such actions will have far-reaching and unwelcome side effects.

Community structure and energy flow

A different approach to the functioning of communities involves attempts to analyse the flow of energy through them (Phillipson 1966). For this purpose all the constituent organisms are arranged in broad categories according to the method they use to obtain nourishment. Thus green plants which obtain nourishment in a single energy step by harnessing light energy are allocated to the first *trophic* or feeding level and are known as primary producers. Plant feeders or herbivores make up the second trophic level and are referred to as primary consumers. Carnivores which eat herbivores are known as secondary consumers, and carnivores which eat other carnivores are tertiary consumers. If the amount of living material produced by the organisms in each trophic level is represented by a series of bars, the diagram which results usually has a pyramidal shape (Fig. 1.6). This configuration arises from the fact that energy entering the system via green plants is lost at each transfer to the next trophic level. It becomes apparent on reflection why this must be the case. No herbivore is able to utilise all its plant food nor any carnivore all its prey. Additionally, a proportion of an organism's food intake is dissipated in respiration and the energy it represents is therefore precluded from passing to the next trophic level. In general terms only

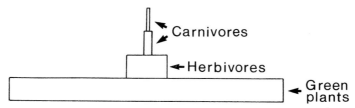

Fig. 1.6. A typical energy pyramid for an aquatic community. For each trophic level the bar widths represent the total amount of energy utilised during the course of a year.

about one tenth of the energy represented by any particular trophic level is passed on to the next one.

This approach to community structure has some obvious practical applications. Strategies for increasing the productivity of agriculture or aquaculture systems are often, in effect, attempts to broaden the plant base of the energy pyramid. It is rarely feasible to increase the supply of light energy to a system; it is often possible, however, to augment the supply of nutrients which would otherwise set a limit on the rate of plant production.

Paradoxically in recreational situations an increase in the productivity of an ecosystem is frequently not regarded as a benefit. Sewage effluents discharged from tourist establishments into lakes are likely to produce the nutrient-rich conditions favourable to dense algal growths. In these circumstances far from serving any useful purpose, such organisms can interfere with recreational activities, both by making conditions unpleasant for swimming, and by interfering with angling by releasing substances poisonous to fish. Moreover, excessive algal growths can have the unwanted effect of stimulating the production of dense swarms of troublesome midges which sometimes render lakeside recreational facilities unusable.

The case histories presented in the chapters which follow should make it abundantly clear that the discipline of ecology, although hitherto neglected in this context, has in fact a crucial role to play in the planning and design of recreational enterprises.

2.
ACTIVE
PHYSICAL
PURSUITS

The range of impacts on biological systems

Recreational activities in which the emphasis is placed on physical skills and endurance can be conveniently grouped together under the heading of 'active physical pursuits'. These include climbing, caving, riding and sailing, and the recreational use of motor-powered devices such as motor boats, snowmobiles, dune buggies and trail bikes. Their potential for damaging biological systems is considerable.

Terrestrial plants can suffer direct damage from the mechanical impacts of trampling feet and vehicle wheels, or may be affected indirectly by modifications to the surrounding soil. These soil changes are the result of compaction in some situations and erosion in others (Liddle 1975). Aquatic plants are vulnerable to the cutting action of boat propellers, mechanical damage from boat hulls, and the erosion of soil from around their roots caused by the wash from moving boats (Liddle & Scorgie 1980).

The reactions of animals to recreational disturbance are more complicated. Some retreat initially from loud noises and unusual visual stimuli, but become reconciled to them if they prove to have no harmful associations. For example, in North America it has been shown that white-tailed deer (*Odocoileus virginianus*) are initially repelled by snowmobiles but after a time will ignore them (Dorrance *et al.* 1975). Unfortunately not all animals can escape direct mechanical injury and the effects of adverse habitat changes. In Florida, injuries

Fig. 2.1. The manatee (*Trichechus manatus*), an aquatic mammal vulnerable to injury from the propellers of power boats.

inflicted by the propellers of power boats represent a major threat to the manatee (*Trichechus manatus*), a docile slow-moving aquatic mammal (Fig. 2.1) (Bertram & Bertram 1973). In the very different habitat represented by subterranean streams, it is the trampling action of cavers and their habit of polluting streams with spent lamp-carbide and batteries which threaten the survival of interesting species of cave crustaceans (Edington & Edington 1977). In terrestrial situations, the pressure of foot and vehicular traffic may not only collapse the burrows of reptiles, mammals and underground-nesting birds, but is also likely to obliterate the interstices normally occupied by microscopic soil animals (Chappell *et al.* 1971). The action of snowmobiles represents a special case of substratum compaction. These machines have been shown to cause high mortality amongst small mammals such as voles, shrews, mice and ground squirrels which spend much of the winter at the junction between the lower snow surface and the ground. The harmful effects probably arise in part from the direct collapse of tunnels, and in part from the altered characteristics of compacted snow which has poor insulating properties. Compaction may also restrict the dispersal of carbon dioxide from respiration and allow the gas to increase to toxic levels (Jarvinen & Schmid 1971).

In the case histories which follow some instances of physical

14

Fig. 2.2. The razorbill (*Alca torda*) (a), and the guillemot (*Uria aalge*) (b), two cliff-nesting birds vulnerable to disturbance by climbers.

pursuits threatening individual plant and animal populations are discussed first. Consideration is then given to situations where recreational disturbances have more far-reaching effects on entire ecosystems.

Disturbance of cliff-nesting birds by climbers

In Western countries cliff-nesting seabirds and their eggs are now largely protected from human exploitation as a source of food. However in some areas the development of sea-cliff climbing as a recreational pursuit has become a threat to these species. Around the coast of Britain, sea cliffs form the major breeding sites for razorbills (*Alca torda*) and guillemots (*Uria aalge*) (Fig. 2.2), the guillemots generally occupying the narrower and more exposed ledges. Sudden disturbance of a breeding colony can cause eggs to be knocked from the ledges as the birds fly off. Even those eggs which survive this fate are liable to become chilled or are attacked by gulls if the return of the parent birds is delayed.

The colonies on small offshore islands have generally escaped disturbance from climbers because of the inaccessibility of such sites. At other localities the risk of disturbance is considerable. This is in spite of the fact that birds and climbers often use substantially different sectors of rock faces. For example, on the cliffs at South Stack in North Wales (Fig. 2.3), the breeding colony of approximately 1500   15

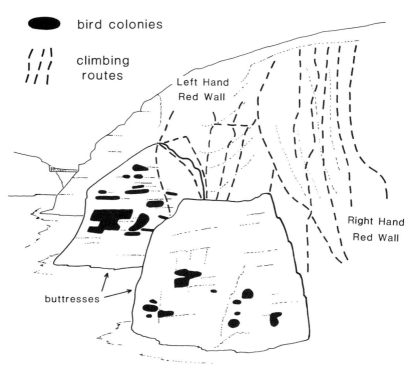

bird colonies

climbing routes

Left Hand Red Wall

Right Hand Red Wall

buttresses

Fig. 2.3. The distribution of guillemot and razorbill breeding colonies on the cliffs at South Stack, North Wales in relation to climbing routes (climbing routes from Sharp 1977).

guillemots and 200 razorbills makes particular use of two large buttresses which project out to sea. On the sides of these buttresses the rock folds have eroded in such a way as to produce a large number of suitable nesting ledges. The buttress sides are of little interest to climbers who favour the main part of the cliff where the rock has fractured across the line of the folds to produce a series of fissured rock faces (Fig. 2.3). Although this would seem to afford a considerable measure of separation between birds and climbers, in practice it has proved to be insufficient to prevent serious disturbance. The starting points of the climbs on the Left Hand Red Wall are reached by abseiling onto the adjacent buttress which in the breeding season is crowded with birds. Additionally, any climber getting into serious

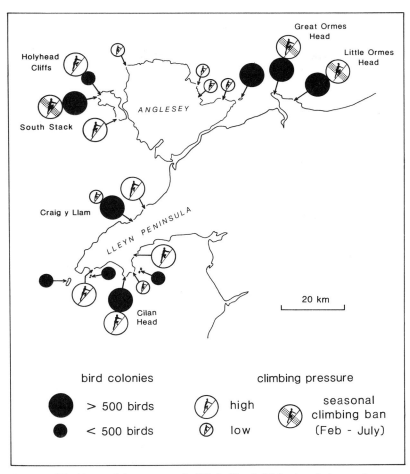

Fig. 2.4. Climbing activity on coastal cliffs in North Wales in relation to the distribution of guillemot and razorbill breeding colonies (bird numbers from Rees 1974).

difficulties on the wall would have to be taken off by helicopter. If such an eventuality arose during the breeding season it would seriously disturb the colony and affect its breeding success.

17

Climbing restrictions

For these reasons, the relevant conservation organisations (the Nature Conservancy Council and the Royal Society for the Protection of Birds) have felt it necessary to negotiate with the British Mountaineering Council for voluntary restrictions on climbing during the breeding season. These apply at South Stack, and on sections of the climbing areas at two other localities in North Wales, Great Ormes Head and Little Ormes Head (Fig. 2.4). A residual difficulty is that not all climbers regard themselves as being bound by voluntary agreements negotiated by the British Mountaineering Council. In any event such agreements need to be kept under review to ensure that they keep pace with changes in the patterns of climbing activity and bird distribution.

Disturbance of waterbirds by sailing boats on reservoirs

Britain has a special part to play in accommodating populations of waterbirds which migrate from Iceland, Scandinavia and the USSR each year to overwinter in Western Europe. Of the birds which visit Britain, up to 25% of some species make use of reservoirs rather than natural water bodies (Atkinson-Willes 1969). These sites remained relatively free from human disturbance until the nineteen sixties when water agencies became persuaded that reservoirs in their care should be made available for public recreation. This resulted in the development of a wide range of activities including angling, sailing, canoeing, rowing, motor-boating, subaqua diving and swimming (Tuite 1982). Whilst such developments have been welcomed by recreational groups, they threaten to jeopardize the role of reservoirs as water bird sanctuaries.

Surprisingly for an ostensibly peaceful pursuit, sailing proves to have a seriously disturbing effect on waterbirds, particularly when it is allowed to extend into the winter period. Batten (1977) describes how the Brent Reservoir in North West London was virtually deserted by teal (*Anas crecca*) and wigeon (*Anas penelope*) in 1963 after winter sailing was permitted in the previously unused northern arm of the reservoir. Similarly, on the Island Barn Reservoir near London, Parr (1974) recorded a substantial reduction in the numbers of overwintering teal after the introduction of winter sailing. Goldeneye

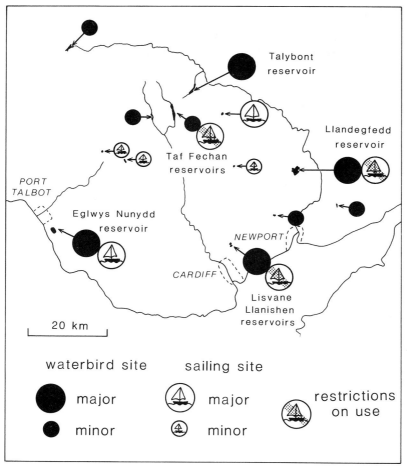

Fig. 2.5. The use for sailing of reservoirs in South Wales, in relation to their importance as overwintering sites for water birds.

(*Bucephala clangula*) are regarded as being particularly sensitive to disturbance, flying up when sailing dinghies approach within 300–400 metres (Hume 1976). Pochard (*Aythya ferina*), tufted duck (*Aythya fuligula*), goosander (*Mergus merganser*), and mallard (*Anas platyrhynchos*) are reputedly less sensitive, tolerating a closer approach of sailing craft before flying up, and returning more readily when sailing finishes at the end of the day.

19

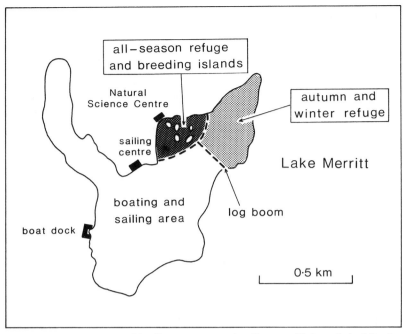

Fig. 2.6. Measures to provide all-season protection for waterbirds at Lake Merritt, City of Oakland, California.

Remedial measures

Such findings have prompted water agencies in Britain to look for ways of managing reservoirs which strike a balance between recreational activities and waterbird conservation. The successful arrangements which are in force in one particular region (South Wales) serve to illustrate the range of the possibilities (Fig. 2.5). The introduction of sailing has been totally resisted at one site (Talybont) and has been limited to the summer months at another (Llandegfedd). At each of the double-basin sites of Taf Fechan and Lisvane/Llanishen, one basin has been designated as a waterbird refuge and the other has been allocated to sailing. Similar concessions to waterbird conservation have yet to be made at the remaining major site in the region, Eglwys Nunydd, an industrial reservoir with a strong and well patronised recreation club.

Summer disturbance of waterbirds

Although most attention has been given to the disturbance of overwintering birds, the requirements of summer breeding populations may also need to be considered. This is often difficult to arrange in the face of peak recreational demands. An instructive example is available from Lake Merritt in California, where the protective measures are adjusted to take account of changing seasonal requirements. In summer, a log boom is positioned to exclude sailing boats from a series of islands used by nesting herons and egrets. In winter when the flocks of ducks arrive, the boom is extended across the north-east arm of the lake to exclude boats from a larger area (Fig. 2.6).

Damage to mountain vegetation by climbers and horseriders

Climbing and arctic-alpine plants in Snowdonia

Turning from animals to plants, mountain habitats provide some particularly interesting examples of damage to plant communities. The mountains of the Snowdonia National Park in North Wales, represent one of the few remaining refuges for arctic-alpine plant species in Britain. These plants were widespread in early post-glacial times but have since become restricted in their distribution as a result of climatic amelioration and stock grazing. The association of the plants with rocky outcrops makes them potentially vulnerable to damage by climbers. This damage may be caused accidentally or as a result of 'gardening' climbs to increase the availability of holds.

Fortuitously, in some parts of the mountain range geological factors intervene to separate the climbers from the plants. The plants tend to be associated with calcium-rich, easily weathered rocks whilst the climbers generally favour the harder, more acidic rock formations. (Fig. 2.7). Unfortunately this separation breaks down in localities where a tradition of climbing on softer rocks has developed. One such example is at the head of Cwm Idwal where climbs have been sited both on the soft basalts and bedded pyroclastic rocks, as well as on

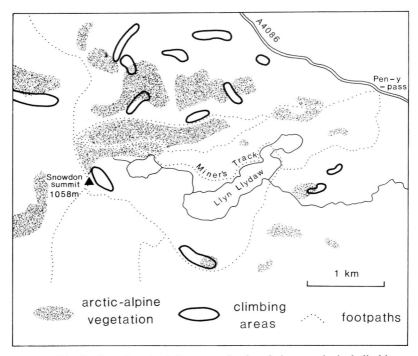

Fig. 2.7. Distribution of arctic-alpine vegetation in relation to principal climbing areas in the vicinity of Snowdon, North Wales Vegetation damage is minimised by the climbers' avoidance of softer rocks (plant distribution from Nature Conservancy Council records).

the hard rhyolitic tuffs (Fig. 2.8). In fact some of the climbs on softer rocks in the Devil's Kitchen are internationally famous. As these cliffs are also of great botanical significance and include one of the last remaining refuges of the rare Snowdon lily (*Lloydia serotina*) a conflict of interest exists which has yet to be resolved.

The impact of riding parties on alpine meadows in North America

In the less-accessible mountain areas of North America, horses and mules are used widely to carry visitors and their camping equipment. Concern has been expressed that this pattern of activity could be

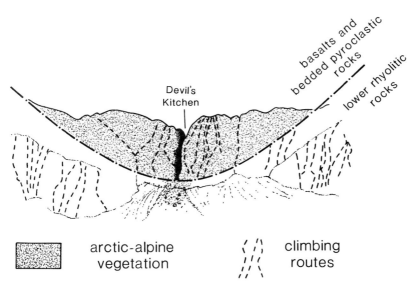

basalts and
bedded pyroclastic
rocks

lower rhyolitic
rocks

Devil's
Kitchen

arctic-alpine
vegetation

climbing
routes

Fig. 2.8. Distribution of arctic-alpine vegetation and climbing routes at Cwm Idwal in the Snowdonia National Park. The climbers' use of softer rocks (basalts and bedded pyroclastics) increases the risk of damage to vegetation (plant distribution from Nature Conservancy Council records, climbing routes from Leppert 1982).

causing damage to natural vegetation, and particularly to alpine meadows. These meadows are a special feature of the Rocky Mountains and the Sierra Nevada range. Some of the characteristic meadow plants are illustrated in Fig. 2.9.

Trail animals can endanger meadow communities by excessive grazing in the vicinity of camp sites, and by loosening the soil on trails so that it washes downslope and buries adjacent vegetation (McQuaid-Cook 1978). Attempts to solve these problems in the National Parks of the Western Mountains have mainly taken the form of restricting the use of trails, campsites and grazing areas in sensitive localities. A typical example is provided by the arrangements made at Rae Lakes Loop, a popular mountain trail in California's Kings Canyon National Park. In the sensitive areas camping is prohibited, or limited to single-night stops, and grazing is controlled on the basis of stock numbers and duration of stay (Fig. 2.10). At some sites animals are

23

Fig. 2.9. Some typical plants from the mountain meadows of the Sierra Nevada:
(*a*) indian paint brush (*Castilleja miniata*); (*b*) broad-leaved lupine (*Lupinus latifolius*); (*c*) shooting star (*Dodecatheon jeffreyi*); (*d*) alpine aster (*Aster alpigenus*);
(*e*) sierra gentian (*Gentianopsis holopetala*).

permitted to graze for only a single night, and an upper limit of 12 animals per party is imposed. At other sites grazing is restricted by allowing only walking parties accompanied by mules. Animal feeding requirements beyond these have to be catered for using bales of hay carried for the purpose. Additionally an overall limit of 30 persons per day is imposed on the trail system as a whole. The regulations

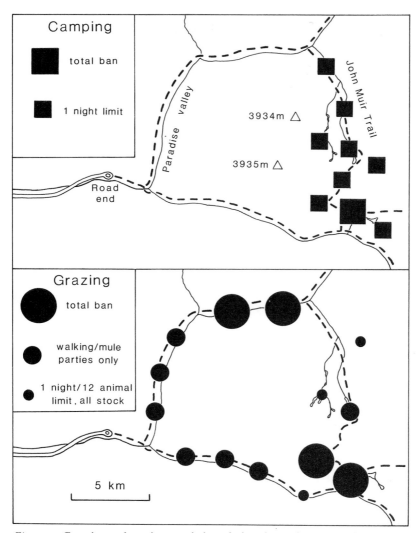

Fig. 2.10. Camping and grazing restrictions designed to reduce vegetation damage on the Rae Lakes Loop trail in the Kings Canyon National Park, California.

are enforced by warden surveillance and by visitors being obliged to submit detailed itineries when applying to the Park Ranger for entry permits. The measures have been generally successful in checking and reversing vegetation damage (Strand 1972, Parsons 1983).

At lower elevations, flower-rich meadows are frequently located in valley sites where waterlogging of the soil has the effect of checking encroachment by forest trees. These sites are vulnerable to the effects of recreational use in a different way. If trails used by riders and walkers are allowed to cut down into the soil surface, this increases the efficiency of drainage and causes drying out. Tree seedlings are then likely to invade the meadow and change its character permanently. To prevent this happening it is now customary to confine trails to the meadow margins.

Damage to desert ecosystems caused by the use of off-road vehicles

When recreational activities involve the use of vehicles in fragile habitats, the harmful effects are frequently not limited to a few species but can result in widespread disruption of whole ecosystems. The operation of off-road vehicles such as trail bikes, dune-buggies and jeeps in desert and coastal sand dune areas provides some particularly striking examples of this kind of disturbance. The potential scale of such activities is illustrated by the 'Hare and Hounds' trail-bike race between Barstow and Las Vegas which in some years has involved more than 2500 riders setting out in a line across the California Desert (Fig. 2.11).

Destruction of desert vegetation

The destruction of shrubs and trees resulting from this single event has been estimated at 140000 creosote bushes (*Larrea divaricata*), 64000 burro-bushes (*Franseria dumosa*) and 15000 Mojave yuccas (*Yucca schidigera*) (Luckenbach 1975). Such damage arises partly as a result of direct mechanical impacts and partly as a result of soil erosion exposing plant roots (Wilshire *et al.* 1978). Many desert areas in California have suffered in this way as a result of recreational vehicle activities and where photographic records are available the progressive reduction in shrub and tree cover is readily apparent.

The other important group of desert plants, the annual ephemerals have also been shown to decline in vehicle recreation areas. It is known that normally the seeds of these plants lie dormant until a succession

Fig. 2.11. Starting line for the 'Hare and Hounds' Barstow to Las Vegas race (photograph by courtesy of the Bureau of Land Management).

of rains causes them to germinate and flower. The rain water stimulates development by washing away chemical inhibitors which are present in the seed covering (Went 1955). The adverse effect of vehicles acts through soil compaction which prevents water penetrating down to the seeds to start the activation process (Luckenbach 1975, Webb 1982).

Harmful effects on animals

The effects of vehicle use on desert animals are both direct and indirect (Fig. 2.12). The animals depend on plants in a variety of ways and are likely to suffer ill-effects from the loss of vegetation cover. For example, the now-endangered desert tortoise (*Gopherus agassizi*) needs to feed on annual plants to build up an energy store for egg-laying and winter survival. Similarly desert rodents depend on the periodic production of seeds from the annuals, and because different species use different sized seeds, any reduction in the diversity of the plants   27

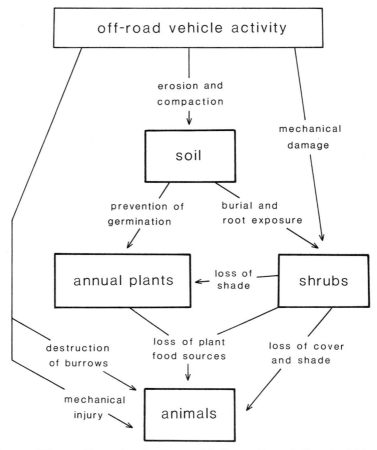

Fig. 2.12. Diagram illustrating the direct and indirect effects of off-road vehicles on the desert ecosystem.

is likely to have repercussions for the rodent populations (Brown & Lieberman 1973). The leopard lizard (*Crotaphytus wislizenii*) uses shrubs for cover, darting out into the open to capture passing insects and other small prey. The brush lizard (*Urosaurus graciosus*) stalks its prey amongst the branches and twigs of shrubs and is well camouflaged for this purpose. The shrubs are also essential to many desert song birds which use them as nesting sites.

Animals can also suffer the ill effects of vehicle use in a more direct

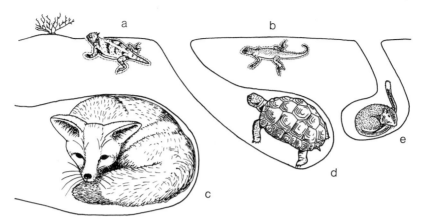

Fig. 2.13. Some typical desert animals in the positions adopted during the heat of the day: (a) desert horned lizard (*Phrynosoma platyrhinos*); (b) fringe-toed lizard (*Uma scoparia*); (c) kit fox (*Vulpes velox*); (d) desert tortoise (*Gopherus agassizi*); (e) kangaroo rat (*Dipodomys deserti*).

way, by being physically injured or by having their burrows collapsed. At any time during the 24–hour cycle there are many more animals present in the desert than are visible on the surface. Lizards, tortoises and snakes need some warmth to become active, but during the hottest part of the day most have to bury themselves in the sand or retreat into hiding to avoid heat damage (Fig. 2.13a, b, d). They tend to adopt the same strategy when falling night-time temperatures begin to make them torpid. Mammals, such as kangaroo rats (*Dipodomys spp.*) and the kit fox (*Vulpes velox*), are in a different position. Being able to regulate their body temperatures, they have no restriction on night-time activity and are able to search for their food at night but remain in their burrows during the heat of the day (Fig. 2.13c, e).

In these circumstances the numbers of animals which can be found crushed on the ground after the passage of recreational vehicles, although substantial enough, are likely only to represent a fraction of the total mortality (Stebbins 1974; Luckenbach 1975). Many more must perish in burrows, which are often sufficiently fragile to be collapsed by foot pressure, let alone the weight of a vehicle.

In an attempt to obtain an overall picture of animal losses from both direct and indirect causes, comparisons have been made of mammal, 29

Fig. 2.14. Bureau of Land Management sign indicating the closure of an area to off-road vehicles except on existing roads and trails.

reptile and bird populations at sites subjected to different levels of recreational pressure (Bury *et al.* 1977). It was found that there were 45% fewer mammals and reptiles at heavily used sites, and in the 'pit' areas used as bases by vehicle operators the reduction was as great as 80%. Similar appreciable reductions have also been demonstrated in bird populations (Luckenbach 1978).

Control of off-road vehicles in desert areas

Against this background, there has been increasing pressure for greater restrictions on off-road vehicle use in desert areas. This has stimulated strong counter-representations from vehicle users. In California the Bureau of Land Management has devised a zoning plan for the 12 million acres of desert land which come under its jurisdiction (Stebbins 1974). In this plan, produced in 1973, vehicles were allowed access to existing cross-country courses and competitive event sites, and to designated vehicle-recreation zones. Some of these latter areas were made available without restriction, others had conditions attached to their use. Additionally vehicles were allowed on all 'existing' trails on land administered by the Bureau. Conservation

30

measures in the plan provided for the closure of sixteen small areas to vehicle use, although some of these had an established trail passing through them.

The 1973 plan was criticised for giving too little attention to the needs of environmental protection (Carter 1974) and since that time the Bureau has extended its consultation and zoning procedures and has closed further areas to vehicle use (Fig. 2.14) In devising measures to reduce habitat damage the Bureau has rightly stressed the need to distinguish between the competitive use of off-road vehicles, and their role in transporting people to inaccessible places to participate in unobtrusive activities such as hiking and family camping. A larger measure of self-regulation is anticipated amongst this latter group of users.

Recreational vehicle damage to coastal sand dunes

Of all the kinds of habitat damage caused by recreational vehicles, that sustained by certain types of coastal sand dunes must qualify as having the most direct implications for human welfare. The critical areas in this context are the dune lines which serve to check the landward penetration of the sea in storm-prone areas and in so doing may significantly reduce the risks to human life and property. The protective function of the dunes depends to a large extent on the stabilising and binding properties of dune vegetation.

In North America 'dune busting' is one of the major causes of vegetation damage and dune destabilisation. This sport involves attempting to drive a vehicle at the face of the dune and up over the dune crest. A series of such vehicle movements along the same route can destroy all existing vegetation and thereby create a sand surface which is likely to be eroded by the wind to form a progressively deepening gulley (Fig. 2.15b) (Mahoney 1980). In East Coast areas of the United States where hurricane-force winds can generate storm surges as high as four metres above the normal high tide level, such breaks often represent the weak points through which the sea penetrates the dune line.

Experiments at Cape Cod, Massachusetts, have shown that the stabilising action of vegetation in other parts of the dune system can also be disrupted by vehicle use (Godfrey & Godfrey 1980). In front

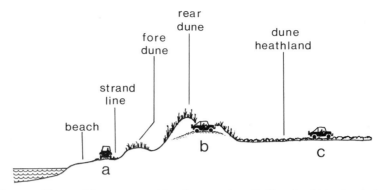

Fig. 2.15. Types of damage caused by the operation of off-road vehicles in sand dune systems: (a) destruction of colonising vegetation along the strand line; (b) creation of tracks across the dune crest which become deepened by wind erosion; (c) destruction of heathland vegetation on the stabilised dunes.

of the dunes, along the strandline, accumulations of rotting seaweed normally provide a useful mulch for colonising grasses. If vehicles are permitted to operate here, this compost-like material becomes dispersed and the grass shoots which it was nurturing are destroyed (Fig. 2.15a). Similarly on the heathland behind the main dunes, vehicle use can re-expose areas which had become completely colonised by vegetation (Fig. 2.15c).

The general conclusion to be reached from studies of sand dune disturbance is that, where a pressing case can be made for vehicular use of the shore, access should be via a limited number of surfaced routes across the dunes, and vehicle activity should be limited to the main part of the beach below the strand line.

Active physical pursuits in perspective

The examples described demonstrate clearly that physical pursuits can damage not only individual plant and animal populations but also whole ecosystems. There are persuasive scientific, aesthetic and practical reasons for seeking to reduce the scale of this damage. Sometimes quite fortuitously the intensity of conflicts is reduced by differences in the habitat requirements of wild species and recreational groups. In other circumstances active intervention is necessary to zone recreational activities or to limit their intensity in the interests of environmental protection. However, whether these measures are formulated on a voluntary basis or imposed by a planning authority, they need to be based on a proper understanding of the ecological processes involved.

3.
OBSERVING
WILDLIFE

In contrast to the physical pursuits described in the previous chapter, there are other recreational activities in which the involvement of wildlife is central and intentional rather than merely accidental. Both casual watching and detailed amateur studies of wild species are activities which come into this category. The increasing popularity of such interests is reflected in the extensive television coverage of natural history topics, the rapidly rising membership of wildlife societies and the development of holidays based principally on animal watching. Undoubtedly the degree of attention given to wild species has had beneficial effects on their welfare by stimulating conservation measures at many levels. In the administration of a number of National Parks for example, the revenue from tourism is more than sufficient to meet the costs of management (IUCN 1976, Budowski 1976). The existence of these and other mutually supportive relationships should not, however, be allowed to obscure the fact that wildlife-based recreation can sometimes produce adverse side effects. Evidence is now accumulating to show that animal watchers can represent a serious source of disturbance even in protected areas.

Some of these adverse effects are obvious and have attracted widespread comment. For example, problems can arise in connection with bird-watching because of competition between participants to accumulate the longest personal tally of species observed. Reports of a rare species are liable to produce an influx of hundreds or even thousands of observers to a site, each intent on seeing the bird in

Fig. 3.1. Bird watchers congregating to observe a rare migrant at Kenfig Local Nature Reserve in South Wales (photograph by courtesy of the Western Mail).

question and if possible photographing it at close quarters (Fig. 3.1). Frequently on these occasions little attention is given to the welfare of the animal itself or to the damage which might be caused to the habitat (British Birds 1982). This is an obvious example of the disturbances which can be caused by animal watchers. In other situations, as will be seen, more subtle disruptions of normal behaviour and ecological relationships are involved.

Distortion of normal behaviour by artificial feeding

Offering titbits of food is a method widely used by the public for bringing animals into closer contact. Whilst in the suburban park or garden this practice does little harm, in more natural settings it can have unfortunate consequences. This is seen most dramatically with larger and bolder species which, having become accustomed to obtaining food in this way, are liable to beg for it in an aggressive fashion. If this behaviour results in human injuries, the offending species frequently suffers by becoming the target of vigorous control measures.

35

Fig. 3.2. Species likely to develop aggressive begging traits if routinely offered food by visitors at recreational sites: (a) racoon (*Procyon lotor*), North America; (b) olive baboon (*Papio anubis*), Africa; (c) long-tailed macaque (*Macaca fascicularis*), Asia.

The management problems caused by the habituation of bears to human foodstuffs in North American recreational areas are well known and are discussed fully in Chapter 8 (p. 162). Other species which become similarly aggressive when fed artificially include mule deer (*Odocoileus hemionus*), bighorn sheep (*Ovis canadensis*) and racoons (*Procyon lotor*). In the case of the racoon (Fig. 3.2a) there is the added hazard that a bite from the animal might lead to the transmission of rabies (McLean 1975). There may also be a disease risk associated with attracting ground squirrels (*Citellus* species) into close contact by offering titbits of food. In the western United States these animals serve as reservoirs for the bacteria causing sylvatic plague (*Yersinia pestis*) and the disease can be passed to man by the transfer of infected fleas.

In South East Asia, the long-tailed macaque (*Macaca fascicularis*) (Fig. 3.2c) is a species which readily becomes habituated to human foodstuffs. Macaque troupes in public parks in Malaysia, Singapore and Hong Kong frequently beg for food from visitors. Spencer (1975, 1979) has shown that this habit is accompanied by fundamental changes in the social behaviour of the animals and an increase in aggressiveness. Visitor-orientated macaques become ill-tempered if food is not forthcoming and in the Hong Kong Country Parks attacks on humans annually cause 20–30 injuries of sufficient severity to require hospital treatment.

In Africa, baboons (*Papio* species) (Fig. 3.2b) and vervet monkeys (*Cercopithecus aethiops*) readily develop troublesome begging habits if offered food by visitors. Difficulties have also arisen in connection with elephants. During the 1950s in the Murchison Falls (Now Kabelega) National Park in Uganda, a large male elephant which had become accustomed to receiving bananas from tourists developed the habit of overturning parked cars apparently to look for food stored underneath, and also took to shaking cars violently to burst open the door- and boot-locks to search for food inside. The fact that people were present in some of the cars did not deter the animal and it was eventually felt necessary to destroy it in the interests of public safety (Grzimek 1964). More recently in a Game Reserve in Zimbabwe, elephants fed with oranges by tourists have adopted the habit of uprooting tents in search of these and other titbits.

The basic solution to these problems lies in convincing visitors that fostering such feeding relationships is not in their own long-term interests nor in those of the wild species they seek to attract. Methods of conveying this message include notices at recreational sites, advisory leaflets and references to the problem in wardens' introductory talks.

Disturbance of relationships with other species

Increased vulnerability to competitors and natural enemies

Observers can also seriously disrupt the life of the species they are watching by unwittingly making them more vulnerable to competition from other species, and to attacks from natural enemies.

SEA BIRDS. At Punto Tombo, a Nature Reserve in northern Patagonia, the entry of visitors into the breeding colonies of king shags (*Phalacrocorax albiventer*) and magellanic penguins (*Spheriscus magellanicus*) has been shown to increase significantly the loss of eggs to predatory gulls (Kury & Gochfield 1975). As a visitor moves into the shag colony the parent birds in the first ten rows or so leave their

a

Western
Gull

Brown Pelican

Raven

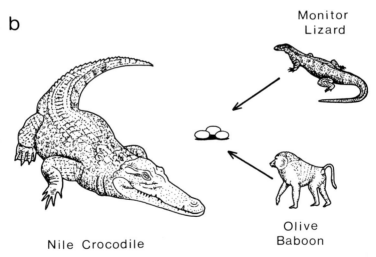

b

Monitor
Lizard

Olive
Baboon

Nile Crocodile

Fig. 3.3. Situations in which the losses of eggs or young to predators are increased by tourist-induced disturbances: (a) predation by western gulls (*Larus occidentalis*) and ravens (*Corvus corax*) on the eggs and young of the brown pelican (*Pelecanus occidentalis*); (b) predation by monitor lizards (*Varanus niloticus*) and olive baboons (*Papio anubis*) on the eggs of the Nile crocodile (*Crocodilus niloticus*).

Table 3.1. *The effect of human disturbance on the breeding success of brown pelicans in the Gulf of California, Mexico (from Anderson & Keith, 1980)*

Year	Number of nests examined	Mean number of young fledged per nest	Decrease of breeding success at disturbed sites (%)
Undisturbed sites			
1971	2556	1.50	
1972	375	1.30	
1974	397	1.19	
Disturbed sites			
1971	965	0.16	89
1972	1609	0.62	52
1974	89	0	100

nests and the gulls, which are constantly patrolling the colony edge, move in to steal the eggs. The gulls seem to recognise the fact that the presence of human visitors creates an opportunity to attack the nests. In Mexico, disturbance by visitors has similarly been shown to reduce breeding success in colonies of the brown pelican (*Pelecanus occidentalis*) (Fig. 3.3a). In this case, the predators are ravens and gulls which exploit the colony by eating eggs and young chicks, and by harassing older chicks until they regurgitate their previous meals (Anderson & Keith 1980). Opportunities for this kind of exploitation are greatest at sites disturbed by visitors, and this fact is reflected in the greatly reduced breeding success of pelican colonies in such situations (Table 3.1). Tourists can aggravate the problem indirectly by discarding edible refuse which, by providing an extra food source, further stimulates the increase of gull populations (p. 178).

CROCODILES. Amongst reptiles, the Nile crocodile (*Crocodilus niloticus*) (Fig. 3.3.b) provides a noteworthy example of a species whose vulnerability to predator attack is increased by the presence of tourists. When crocodiles were a major tourist attraction in the Murchison Falls National Park in Uganda regular boat trips were organised to view the riverbank breeding areas. The females actively protect their eggs, which are buried in the sand for the three-month

Table 3.2. *The effect of visits by tourists*
on the destruction of crocodile nests by predators.
Data from sites in Uganda (from Cott 1969)

Breeding ground	Number of nests	Number of nests destroyed by predators (%)
Sites not visited by tourists		
Falls Bay	36	17 (47)
Frog Bay	7	2 (29)
Sand-river 2	22	4 (18)
Paraa South	7	0 (0)
Tourist sites		
Sand-river 4	13	13 (100)
Namsika	10	10 (100)
Bee-eater slope	13	7 (54)

incubation period, and also defend their young after hatching. The arrival of tourist boats caused the females to retreat from the bank into the water and this allowed predators, particularly monitor lizards (*Varanus niloticus*) and olive baboons (*Papio anubis*), to move in and take eggs or young (Cott 1969). A monitor lizard was seen to devour a dozen eggs in 78 minutes and baboons often departed carrying three or four eggs each. In consequence, losses to predators were significantly greater at localities visited by tourists than at those which remained free from this kind of disturbance (Table 3.2).

CHEETAHS. Accounts in local natural history journals in East Africa suggest that amongst the mammals, cheetahs (*Acinonyx jubatus*) may also suffer from tourist-induced disturbances. One observation from the Amboseli National Park describes how a tourist minibus drawing up alongside a previously hidden cheetah and her young, appeared to attract first a flock of vultures and subsequently a group of lionesses to the site. As a result the cheetah family was obliged to flee and abandon the antelope carcase on which it had been feeding. Similar incidents have been recorded from the Serengeti National Park which suggest that hyenas similarly use stationary minibuses as a means of locating and robbing cheetah families of their prey. Apparently they sometimes kill the cheetah cubs in the process.

Fig. 3.4. A Galápagos land iguana (*Conolophus subcristatus*).

Disturbance of relationships between members of the same species

Interference with territorial behaviour

For many animals, establishing a territory is part of the essential process of securing a mate, and of ensuring access to adequate food resources. Tourist activities can interfere with these territorial mechanisms in some unexpected ways.

In East Africa, it has been found that female Thomson's gazelles (*Gazella thomsoni*) are more timid than the males and readily leave the breeding territory when disturbed by the approach of a tourist vehicle. The net result of this behaviour is to separate the sexes for long periods with an adverse effect on breeding success (Walther 1969).

Another example of territorial disruption has been observed amongst the land iguanas (*Conolophus subcristatus*) (Fig. 3.4) on the now much-visited Galápagos Islands. The male lizards normally maintain strictly separate breeding territories. It was discovered, however, that on the island of South Plaza, this territorial system had broken down because the animals were leaving their territories to congregate in areas where they could beg for titbits from visitors. As a result, breeding activity had come to a standstill (Harris 1973). In this instance a strictly enforced ban on artificial feeding was sufficient to resolve the problem.

Ecology, recreation and tourism

Fig. 3.5. A female wildebeest (*Connochaetes taurinus*) and her young calf. Disturbances which cause the temporary separation of newly born animals from their mothers can seriously interfere with the formation of mutual recognition bonds (photograph – Camerapix Hutchinson Library Ltd).

Disruption of bonds between parents and offspring

HOOFED ANIMALS. The movement of tourist vehicles can have the effect of increasing the mortality of young animals by separating them from their mothers during the critical period when recognition bonds are being established. Recent animal behaviour studies (Lent 1974) have shown that in many hoofed animals (ungulates) there is a critical 'bonding period' immediately after the birth during which the mother learns to recognise her offspring individually. This recognition system is probably based on taste and odour characteristics perceived initially by the mother when she licks the newborn infant, and subsequently reinforced by differences in voice and coat patterns (Fig. 3.5). The time taken to establish such bonds varies from species to species, but ranges from five minutes to six hours. After the bonding period, some species hide their young in dense vegetation and return periodically to suckle them. Others maintain more continuous contact with their offspring until they are weaned. The first group of 'hider' species includes most deer and antelope, whereas the second group of

42

'follower' species is exemplified by sheep, chamois and wildebeest. Whichever pattern applies, the establishment of the initial mother/offspring bond seems to be essential for successful rearing, because it is on this basis that the mother accepts the young for suckling, attempts to protect it from predators, and in the case of the 'hider' species returns to it in its hiding place. After weaning there is a general decline in the strength of maternal-infant bonds and a progressive tendency for the growing animal to make wider social contacts within its species group.

In East Africa, many of the more perceptive visitors to National Parks have expressed concern that carelessly operated tourist vehicles frequently separate young animals such as zebra and wildebeest from their parents (Lawick-Goodall & Lawick-Goodall 1970). Such disquiet is justified. Separation is potentially dangerous for the young animals at any stage, but is especially serious if it occurs before the bonding process is complete. In these circumstances even if the offspring is physically reunited with the mother it is unlikely to be accepted and will almost certainly die or be killed by predators. Separation after the bonding period is not necessarily fatal for the young because if the bonds have been properly formed, mothers can identify their offspring, and search for them. However, the more the herd is scattered by a disturbance, the greater are the chances of attack by predators and the slimmer the chances of mother and offspring being reunited.

WHALES. Similar concern about the disruption of parent/offspring bonds has recently been expressed in relation to 'whale watching'. This activity has formed the basis for a rapidly expanding tourist industry and whale-watching cruises are now available along the Pacific and Atlantic coasts of North America, and in the seas around Hawaii, the West Indies, Sri Lanka and New Zealand (McCloskey 1983). The scale of such operations can be gauged from the fact that an estimated one million people have taken whale-watching cruises from Californian ports since 1968 and that the annual income from this activity is in the region of 1.5 million pounds sterling (Kaza 1982).

Whale watching along the west coast of North America is based mainly on the migrations of the gray whale (*Eschrichtius robustus*) which moves annually between summer feeding regions in the Arctic seas to calving areas along the Mexican coast (Fig. 3.6). Whale

43

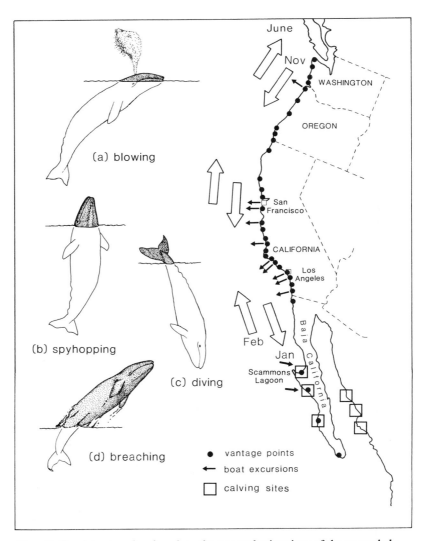

Fig. 3.6. Tourist enterprises based on the seasonal migrations of the gray whale (*Eschrichtius robustus*) along the west coast of North America.

watchers frequently have the gratification of observing spectacular manoeuvres by these 40-ton marine animals. These include blowing stale air from the lungs, diving, 'spyhopping' in which the whale protrudes its head from the water, and 'breaching' when it rises out of the water and drops back with a tremendous splash (Fig. 3.6a–d). The purpose of these last two activities is still not understood.

Whale calves normally maintain constant bodily contact with their mothers but when separated experimentally are likely to transfer their attachment to the side of a nearby ship (Norris *et al.* 1977). This fact has given rise to speculation that the manoeuvring of whale-watching boats could cause such transfers to occur accidentally and result in the irreversible separation of calves from their mothers. It is also likely that the noise made by boat engines and propellers interferes with the sound-communication systems used by the animals (Payne & McVay 1971).

The design of control measures to reduce disturbances associated with animal watching

Zoning

An obvious method of preventing unnecessary disturbance to wildlife populations is to bar public access to them at times when they are most vulnerable. This principle is employed in a number of zoning systems. For example in the Great Barrier Reef Marine Park in Australia, four out of the fourteen reef islands in the Park are closed to public access between October and March each year to prevent disturbance to nesting birds and turtles. During this period there is also a ban on overflying by aircraft at altitudes below 1000 ft. A further island (Wreck Island) is closed at all times to public access with the aim of safeguarding its biological systems from any kind of alteration likely to result from human use (Great Barrier Reef Marine Park Authority 1980).

With land-based National Parks, there is similar scope for designating 'strict natural areas' to which public access is prohibited or severely restricted (Forster 1973). For example, the plan for the Bialowieza National Park in Poland involves a system of three concentric zones in which the central core area is reserved for

45

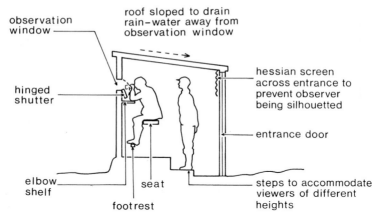

Fig. 3.7. Design features which can be used to increase the effectiveness of animal-watching hides (based on Beale & Wright 1968, and Beazley 1969).

scientists with special permits or visitors under the strict supervision of guides. The more remote and inaccessible areas of parks lend themselves to special designations of this kind.

The use of fixed viewing points

Another solution to disturbance problems is to confine watchers to fixed observation points sited at safe distances from those parts of the habitat where wildlife is likely to appear. Viewing hides constructed for this purpose are now a familiar feature of many nature reserves and national parks, and increasing attention is being given to the incorporation of design features to make these facilities more effective (Fig. 3.7). Frequently there is also scope for modifying the surrounding habitat to make it attractive to animals. At some sites this has involved the construction of artificial pools and waterholes, at other sites alterations have been made to the vegetation. An example of this latter approach is seen in West Malaysia's National Park, Taman Negara, where forage grasses have been planted in mineral-rich seepage areas ('saltlicks') adjacent to observation hides (Fig. 3.8). The improved grazing provides an additional attraction for species such as tapirs (*Tapirus indicus*) and wild cattle (*Bos gaurus*).

Undoubtedly the most elaborate animal-watching facilities are

Fig. 3.8. An observation hide sited alongside a 'saltlick' in West Malaysia's National Park, Taman Negara (note the screening designed to hide the access ladder).

represented by the viewing lodges in East Africa. At Salt Lick Lodge in Kenya visitors are accommodated in individual rooms in elevated hut-like structures which serve both as bedrooms and observation points. From these rooms and from the verandah of the restaurant it is possible to watch elephants, warthogs, antelopes and zebras as they visit the nearby waterholes (Fig. 3.9). A closer view of one of these waterholes can be obtained by following an underground passage to a sunken observation chamber. Similar animal-watching facilities are available in the longer-established viewing lodges such as Treetops in the Aberdare National Park.

It is not always appreciated that a considerable amount of 'stage management' may be necessary at these centres to ensure that visitors enjoy good views of the animals. At Salt Lick Lodge one of the waterholes is in fact a concrete-lined structure filled by a piped water supply from a dammed lake on a nearby river. At Treetops the supply to the waterhole is similarly supplemented to prevent it from drying out. At this centre other devices used to improve the visitors' prospects of seeing animals include the sprinkling of salt around the lodge to act as an attractant and the enclosure of the surrounding park

47

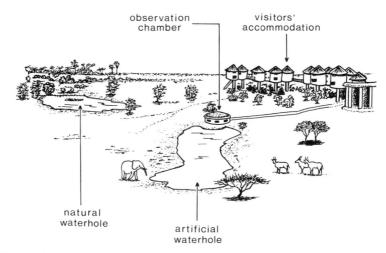

observation
chamber

visitors'
accommodation

natural
waterhole

artificial
waterhole

Fig. 3.9. Animal-viewing arrangements at Salt Lick Lodge in Kenya.

within an electric fence and ditch to prevent animals dispersing into the surrounding countryside (Cherfas 1984). If such arrangements can be made unobtrusively they need not detract significantly from the feeling of 'naturalness' enjoyed by visitors.

The additional possibility of using food lures to attract animals to viewing sites must clearly be regarded with caution in the light of the examples of food habituation described earlier (p. 35). However, in the East African parks the provision of carcases by park staff to attract lions or leopards probably involves little disruption of the normal activities of these animals. Lions normally scavenge a proportion of their kills from other species such as hyenas, wild dogs, cheetahs and leopards (Schaller 1972) and it probably makes little difference whether the carcase used has been left by another predator or procured by the game wardens. The practice which has been used in the Gir National Park in India of allowing lions to attack tethered buffaloes as a tourist spectacle is inhumane and must be regarded as unacceptable.

The criticism that even the best viewing facilities involve a large element of artificiality must be balanced against the difficulties of providing a rewarding wildlife experience for a sizeable number of people by any other means.

Guidelines for viewing mobile subjects

The problems of control multiply in situations where animals can be viewed satisfactorily only if the observers are allowed a considerable measure of mobility. Occasionally an inspired solution presents itself, as in India's Kaziranga National Park and at Chitawan National Park in Nepal. Here riding elephants are being used successfully to allow visitors to view the Indian rhinoceros (*Rhinoceros unicornis*) at close quarters (O'Connor 1980). Elephants have the advantage that they provide a high viewing point and are regarded with relative equanimity by the rhinos.

The problem of exerting adequate control over tour minibuses in the East African parks is a much more difficult one. Competition between tour operators is intense and the guides are frequently under pressure from their passengers to seek out the more unusual carnivores such as leopards and cheetahs for close-up observation. Although theoretically it would be possible to define safe-approach distances these would be extremely difficult to enforce. In a few instances strategically placed ditches prevent vehicles from leaving the main track.

The most notable attempt to tackle the safe-distance problem comes not from the land but from the sea, and concerns the regulations relating to whale watching in United States waters. In these sea areas, there are general provisions under the Marine Mammal Protection Act of 1972 and the Endangered Species Act of 1973 which make it illegal to harass, harm, pursue or wound marine mammal species. Within this framework, the National Marine Fisheries Service has drawn up guidelines for watching gray whales off the Californian coast and humpback whales (*Megaptera novaeangliae*) in Hawaiian waters. In the case of the gray whales, the guidelines specify that vessels should not approach whales closer than 100 yards; that in the vicinity of whales, speed should be kept constant, not exceeding that of the slowest whales in a particular group, and that vessels should never separate a whale from its calf. The guidelines for watching humpbacks in Hawaii are generally more stringent in recognition of the greater vulnerability of this species. In Glacier Bay, Alaska, special regulations are in force to reduce disturbance to humpback whales. These limit the total number of vessels entering the Bay and require them to remain at least a quarter of a mile from the animals. The restrictions were prompted by strong circumstantial evidence that increased disturbance by tourist boats was responsible for the avoidance of the Bay by humpbacks in 1978 and 1979 (Marine Mammal Commission 1982).

In our increasingly urban and mechanised society, animal-watching and similar pursuits must be recognised as serving the vital function of allowing people to come into contact with the natural world. However, if these benefits are to be enjoyed without jeopardising the welfare of the wild species involved, constant attention needs to be given to behavioural and ecological factors.

4.
RECREATIONAL
HUNTING
AND
FISHING

In spite of claims to the contrary it is clear that recreational hunting and fishing can represent a major source of environmental disruption. It is not necessarily the species which are actually being hunted that are in jeopardy, after all sportsmen have a vested interest in maintaining these populations. The species at risk are more usually those affected by other actions taken to increase the availability of game. The disruption of indigenous populations by the introduction of exotic species, and the routine persecution of predators provide notable examples of disturbing influences of this kind. Additionally in recent years it has become apparent that hunting and fishing can represent a serious source of environmental contamination, because of the accumulation of discarded angling weights and spent lead shot in aquatic habitats.

Hunting in historical perspective

Many early human societies which depended on hunting and gathering activities for food remained in balance with the species they exploited. Whilst this balance apparently stemmed in part from a lack of sophisticated hunting weapons, in some cases it seems to have been associated with a conscious frugality in the exploitation of resources. Similar care for resources has been observed amongst the few aboriginal hunting societies which have survived intact until modern times (Meggitt 1964, Lee 1972). For example, some Australian

51

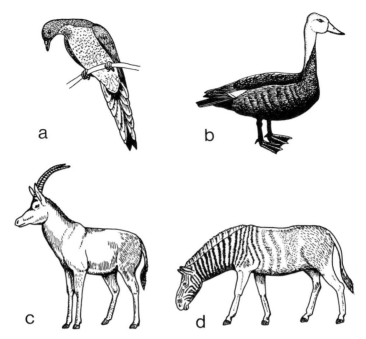

Fig. 4.1. Species whose extinction was accelerated by unrestricted hunting: (a) passenger pigeon (*Ectopistes migratorius*): (b) pink-headed duck (*Rhodonessa caryophyllacea*); (c) bluebuck (*Hippotragus leucophaeus*); (d) quagga (*Equus quagga*). (c and d after Dorst 1970).

aboriginal groups are known to have refrained from hunting stingrays during the breeding season and to have spared female wild dogs to allow the animals to breed again.

There are, however, many striking examples of catastrophic over-exploitation of wildlife populations during historic times. In North America during the last century, hunting by European colonists reduced two bird species, the heath hen (*Tympanchus cupido*) and the passenger pigeon (*Ectopistes migratorius*) (Fig. 4.1a), to extinction, and was responsible for the spectacular decline of a number of others including the whooping crane (*Grus americana*) and the Eskimo curlew (*Numenius borealis*) (Halliday 1978). Dramatic reductions were also brought about in populations of large mammals such as the bison (*Bison bison*) and the pronghorn antelope (*Antilocapra americana*).

52

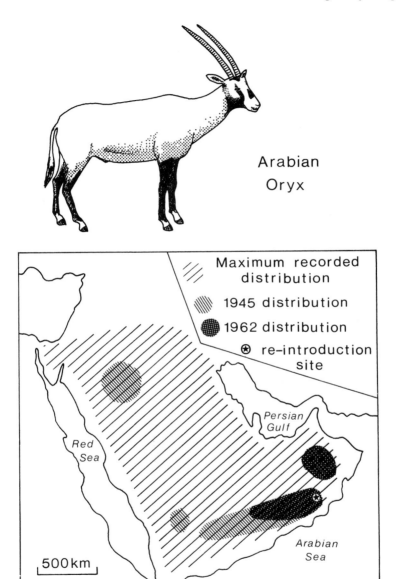

Fig. 4.2. Map showing the progressive reduction in range of the Arabian oryx *Oryx leucoryx*) resulting principally from the effects of uncontrolled hunting (after Stewart 1963).

53

During the same period, the Dutch in South Africa eliminated the bluebuck (*Hippotragus leucophaeus*) and the zebra-like quagga (*Equus quagga*), and the British in India hunted the pink-headed duck (*Rhodonessa caryophyllacea*) to probable extinction (Fig. 4.1b–d). There were various motives behind these activities. In the case of the African and American animals, obtaining meat and skins, and removing indigenous animals from land intended for other purposes were major considerations. The reasons for hunting the pink-headed duck appear to have been purely recreational (Ali 1960).

THE ARABIAN ORYX. A more recent example of the harmful effects of recreational hunting is provided by the near-extermination of the Arabian oryx (*Oryx leucoryx*) in the Middle East (Fig. 4.2). Traditional hunting methods employed by Bedouin tribesmen probably had little effect on the population of this species (Stewart 1963). The situation was transformed, however, by the influx into the region of vehicles associated with the oil industry, and the spread of modern weapons during the Second World War. This made it possible for hunting parties to be organised on a massive and destructive scale. Talbot (1960) describes how hunting groups involving hundreds of vehicles were deployed across the desert to pursue and shoot oryx. These activities have progressively reduced the range of the species to the point where it is now doubtful whether any herds survive, other than those which have been reintroduced recently from captivity (Jungius 1978).

These, and other examples, demonstrate that hunting using modern technology can have disastrous effects on wildlife populations. Need this necessarily be the case?

The scope for exploiting game animals on a sustained-yield basis

Modern recreational hunters make the claim that animal populations can be hunted on a sustained-yield basis, without progressively diminishing their numbers, if the animals are kept free from disturbance during the breeding season, and if the level of harvesting is adjusted to take account of year-to-year variations in reproductive success and winter mortality.

That it is possible to meet these requirements is demonstrated by

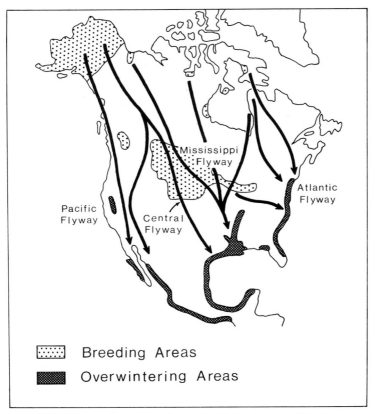

Fig. 4.3. Autumn migration routes of ducks in North America. Hunting regulations take into account year-to-year variations in breeding success of the species in each flyway.

the organisation of duck shooting in North America. The ducks breed mainly in the northern half of the continent and migrate annually to overwintering areas in the south. These movements take place along four main routes or 'flyways' (Fig. 4.3). Shooting restrictions protect the birds during the breeding season, and the level of harvesting during the autumn and winter is adjusted to take account of year-to-year variations in breeding success. For many species reproductive output is closely linked with rainfall in the so called 'duck factory' area of the prairie provinces. This region contains innumerable small ponds or 'potholes', formed originally by glacial action. In wet years many more of these fill with water and provide suitable breeding conditions. Account is taken of these year-to-year variations and their effect on breeding output when framing the hunting regulations for the following winter hunting season. In a poor breeding year, arrangements are made to reduce hunting intensity by shortening the length of the hunting season and reducing the maximum number of birds (the bag limit) a hunter is allowed to take each day. Conversely in a good breeding year the length of the hunting season and the bag limits for the species in question are increased.

A factor which also needs to be taken into account when devising hunting regulations is the relationship between natural mortality and that caused by hunters. In species such as the black duck (*Anas rubripes*) there is evidence that the two sources of mortality are independent (Geis 1963). Thus the number of ducks dying from natural causes is not related to the number shot. In the mallard (*Anas platyrhynchos*) the situation appears to be different. Within certain limits, a reduction in the number shot leads to an increased mortality from natural causes, and the total death rate may remain much the same (Anderson & Burnham 1978, Rogers *et al.* 1979). For species in which this latter type of compensatory process occurs, 'fine-tuning' the hunting regulations, although involving much administrative effort, is unlikely to have a significant effect on the overall result. In such circumstances, as was pointed out by the pioneer wildlife biologist J. L. Lynch in the 1950s, rather broad directives on harvesting levels are all that are required (Boyd & Lynch 1984). Unfortunately, few hunting enterprises attempt to match the level of ecological and administrative inputs devoted to the regulation of duck shooting in North America. Consequently a number of them run the risk of de-stabilising the animal populations being hunted.

Control actions against predators

A notion prevalent amongst recreational hunters is that they are competing for quarry species with natural predators. Therefore by controlling these predators, or so it is argued, it should be possible to increase the yield to the hunter. Plausible though this idea is, it gains only limited support from modern ecological studies.

Outside the shooting season, game animals die from a variety of causes which include starvation and parasitism, as well as attacks by predators. The vulnerable individuals are often those which have been forced into marginal habitats by social pressures such as competition for territories. Such animals can be regarded as a kind of 'doomed surplus'. In these circumstances the removal of predation as a cause of mortality may have little effect on the final death rate because the other mortality factors then increase in importance in a compensatory fashion (Murton 1968). Extensive studies of red grouse (*Lagopus lagopus scoticus*) in Scotland provide a striking demonstration of such mechanisms (Jenkins *et al.* 1964, 1967). The grouse set up territories in suitable heather areas and any birds which are surplus to the carrying capacity of the habitat are pushed out into less-favourable grassy areas. These latter birds are the ones which are most likely to die from starvation, nematode worm infections and attacks from birds of prey such as hen harriers (*Circus cyaneus*) and golden eagles (*Aquila chrysaetos*). Consequently it is unlikely that the routine removal of such predators (which is in any case an illegal activity) would significantly increase the numbers of grouse available in the shooting season, these numbers being largely determined by the area of heather suitable for territories.

In North America, Errington (1946, 1967), working on another game bird species, the bobwhite quail (*Colinus virginianus*), has come to similar conclusions. In this species the most vulnerable individuals are those which have been excluded by social pressures from areas of good food supply and vegetation cover. In these circumstances the number of predators present has little influence on the overall mortality of quail, because birds which escape their attentions are in any event likely to succumb to death by starvation. Likewise with such game species as deer and wild sheep in North America, Connolly (1980) has been able to identify a large number of situations where the size of the prey populations appears to be limited by factors other than predation.

Circumstances which increase the vulnerability of game animals to predators

Altogether there appear to be only three sets of rather special circumstances in which game animals become so vulnerable to predators that control measures might reasonably be contemplated.

REINTRODUCED SPECIES. The first of these is where attempts are being made to reintroduce species into the areas they originally occupied. In North America the programme to reintroduce pronghorn antelope (*Antilocapra americana*), mainly for sporting pruposes, into rangeland in Oregon, Arizona, Utah and Texas has been hampered by attacks from coyotes (*Canis latrans*) (Arrington & Edwards 1951). Probably the vulnerability of the antelopes was increased by their unfamilarity with new surroundings. In one instance a herd of seventeen antelope transferred to a site in Oregon in 1969 increased in numbers only after the introduction of measures to control coyotes. Thereafter the herd increased progressively to about 100 by 1976. Similar benefits from control measures have been reported in relation to introductions in the other States.

HIGH WINTER MORTALITY. The second kind of situation in which intervention might be justified is where severe winter conditions have depleted herbivore populations without having a commensurate effect on their predators. In North America this is known to happen occasionally with moose (*Alces alces*) and deer (*Odocoileus hemionus*) populations preyed upon by wolves (*Canis lupus*). The wolves are then likely to concentrate on the remaining survivors and bring about a collapse of both prey and predator populations (Connolly 1980). In these circumstances a case might be made for reducing the wolf populations to stabilise the situation.

GAME-REARING UNITS. Finally there are situations where the rearing of game species in conditions resembling farm units is likely to increase their vulnerability to predator attack. A fox gaining access to a pheasant-rearing pen can cause serious losses, and the large numbers of fish tags sometimes found in the regurgitated pellets of herons are evidence of this bird's ability to exploit fish hatcheries. The phenomenon of 'surplus killing' (Kruuk 1972) may be involved in some of these cases. This occurs when a predator, faced with prey which cannot escape, becomes diverted into an aberrant behaviour

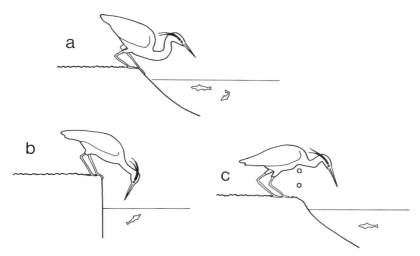

Fig. 4.4. Diagram to illustrate techniques to reduce the losses of hatchery-reared fish to herons. (a) A high water level and gently sloping banks facilitate attacks by herons; (b) lowering the water level and steepening the banks reduce the risks; (c) so also does the erection of cord barriers (after Meyer 1981).

which involves attacking many more individuals than it can possibly use as food.

In most game-rearing operations, however, there is scope for introducing design features which minimise losses due to predators, rather than adopting a policy of predator control. In the case of herons attacking trout in ponds, recent studies have shown that relatively simple modifications to the profile of the bank (making it steeper) or the fixing of cords or floats around the pool edge can reduce dramatically the losses caused by the birds (Fig. 4.4). Likewise in pheasant-release pens, the chances of young birds escaping attack from predators such as owls are greatly increased if a dense cover of shrubs and ground vegetation is provided as shelter.

A rationale for predator management

It must be concluded that an automatic commitment to the control of predators by recreational hunters can no longer be justified. Not only are such practices at variance with our modern knowledge of

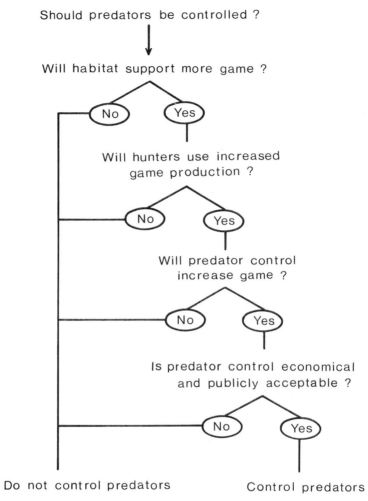

Fig. 4.5. Issues which need to be examined before embarking on a predator–control programme (modified from Connolly 1980).

population dynamics they are also increasingly out of tune with public interest in, and concern for, predatory species. In these circumstances every demand for predator control should be judged strictly on its merits in a proper ecological context. Figure 4.5 provides a possible framework for this kind of decision making.

The disruption of natural communities by introduction of game animals

A further group of problems has been generated by the field sportsman's enthusiasm for introducing exotic game species into new habitats. This practice has disrupted many natural communities and is particularly to be regretted when it disturbs populations essential to our understanding of evolutionary processes.

Relict fish populations in the Death Valley drainage system

One such assemblage is represented by fish populations in the Death Valley region of California and Nevada. Nowadays most of this area is extremely arid with a drainage system represented mainly by desert springs. In late Pleistocene times 10000–20000 years ago, when the climate was wetter the region was characterised by an extensive system of large lakes and rivers (Fig. 4.6a). As conditions became drier, the lakes decreased dramatically in size or disappeared altogether and the streams became intermittent, or reduced to isolated springs (Fig. 4.6b). As a result, fish populations which had formerly been widely distributed became fragmented and began to diverge from one another in isolation. Ten different species or subspecies of pupfish (*Cryprinodon*) and four killifish (*Empetrichthys*) apparently developed in this way (Miller 1950, Brown 1971) (Fig. 4.7a, b). In addition, a more widespread species, the speckled dace (*Rhinichthys osculus*) (Fig. 4.7c), produced different colour variants in different areas. Biologists regard this kind of divergence in isolation as one of the fundamental mechanisms of evolution, and are constantly searching for situations where the process can actually be demonstrated.

It is a source of regret therefore that these interesting fish populations should have been disturbed in various ways by human interference. Of particular relevance in the present context is the disturbance caused by exotic species introduced for sporting purposes (Soltz & Naiman 1978). At Crystal Pool, one of the spring-fed sites in the Ash Meadows system, the introduction of the largemouth bass (*Micropterus salmoides*) (Fig. 4.8a) has caused the extermination of at least three species from the site; the speckled dace, the Ash Meadows pupfish (*Cyprinodon nevadensis mionectes*) and the Ash Meadows

61

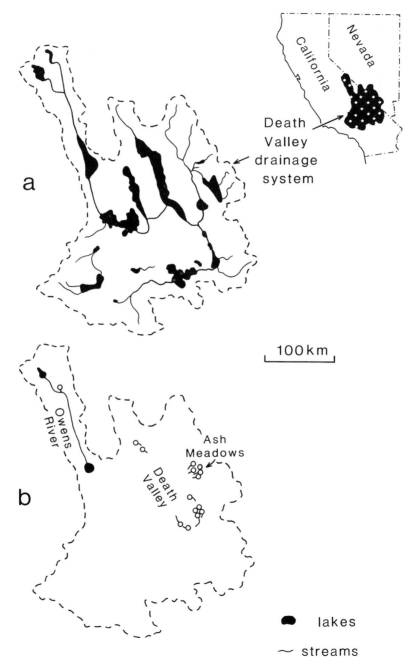

100 km

● lakes

~ streams

○ springs supporting pupfish populations

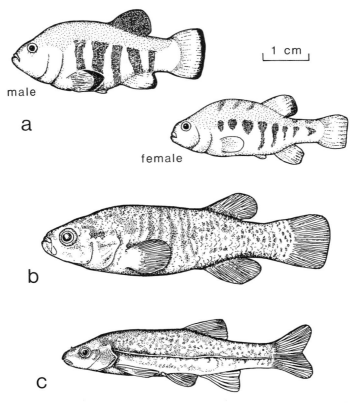

male

a

female

b

c

Fig. 4.7. Indigenous fish from the Death Valley drainage system which have been adversely affected by the introduction of exotic species: (a) pupfish (*Cyprinodon*); (b) killifish (*Empetrichthys*); (c) speckled dace (*Rhinichthys osculus*) (based on Brown 1971, Soltz & Naiman 1978).

killifish (*Empetrichthys merriami*). In fact the Ash Meadows killifish has disappeared from all its previous localities and is now extinct. Similarly in the Owens River system, the introduction of largemouth bass and trout has seriously diminished the populations of both the Owens pupfish (*Cyprinodon radiosus*) and the speckled dace and has

Fig. 4.6. Maps showing: (a) the maximum extent of the Death Valley drainage system in late Pleistocene times; (b) the much-reduced present-day drainage system in the same area and the distribution of relict populations of pupfish (based on Miller 1950, Soltz & Naiman 1978).

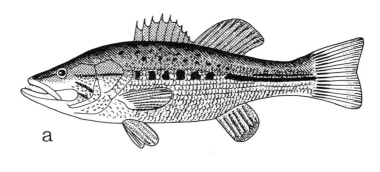

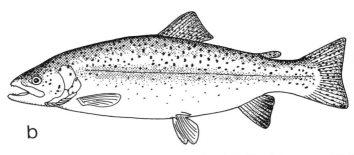

Fig. 4.8. (a) Largemouth bass (*Micropterus salmoides*), (b) rainbow trout (*Salmo gairdneri*): two North American fish whose introduction to new localities has caused disturbance to indigenous species.

caused them to become restricted to a few pools and tributary streams respectively. These disruptive effects of introduced species appear to result either from direct attacks on the indigenous species or from competition for the same resources.

The effects of exotic fish introductions on lake communities

The fish faunas of certain large lakes also provide important demonstrations of evolutionary processes, and unfortunately are equally vulnerable to disturbance by introduced species. Many large lakes appear to have started their biological history with an impoverished fish fauna, but in the absence of competitors these elements have become diversified to produce a wide variety of new species each

64

adapted to utilise a particular habitat. This process has been well documented from some of the Rift Valley lakes in Africa. For example, Lake Malawi and Lake Tanganyika support 'species-flocks' of 200 and 126 species respectively. All but four of these species are known only from these localities (Fryer & Iles 1972).

THE FISH OF LAKE TITICACA. In the high Andean lake of Titicaca (3800 m), there has been a similar ecological radiation of fish types, in this case based on the genus *Orestias* in the carp family (*Cyprinidae*) (Tchernavin 1944, Gilson 1964). These fish have been studied much less than the African species-flocks, but appear to comprise twenty species of various sizes with a variety of feeding habits. Some apparently feed on vegetation, others eat crustaceans, others small molluscs and so on. It is typical of these isolated lake faunas, that members of a single family should diversify to fulfil the range of ecological roles which in a more varied fauna would be distributed amongst a number of families.

Whilst the African lake fish have so far escaped serious disruption by exotic species, this is not the case for Lake Titicaca. Between 1937 and 1942 eggs of the North American rainbow trout (*Salmo gairdneri*) (Fig. 4.8b) were introduced into the lake, apparently with the intention of promoting a recreational fishery (Villwock 1972). Subsequent introductions included brown trout (*Salmo trutta*) from Europe and brook and lake trout (*Salvelinus fontinalis* and *S. namaycush*) from North America. The full impact of these introductions has yet to be investigated. However, there seems to be little question that the largest of the *Orestias* species, *Orestias cuvieri*, has virtually disappeared from the lake. Apart from suffering as the result of competition, this species appears to have been affected by sporozoan parasites introduced accidentally with rainbow trout eggs (Villwock 1972). The presence of the exotic fish now removes for all time the possibility of understanding the habitat relationships of the original fauna.

Interactions between exotic fish and native birds

FLIGHTLESS GREBES. Although in general, it would be expected that the competitive impacts of exotic fish would be felt mainly by other fish, in some instances the ill-effects are experienced by

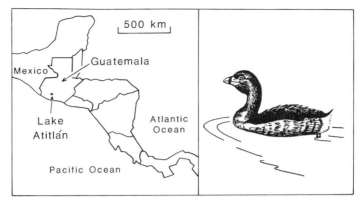

Fig. 4.9. The Giant pied-billed grebe or 'poc' (*Podilymbus gigas*), known only from Lake Atitlán in Guatemala and threatened by competition from largemouth bass introduced for recreational purposes.

unrelated groups such as aquatic birds. A number of the mountain lakes in South and Central America support interesting populations of flightless grebes. One of these is the giant pied-billed grebe (*Podilymbus gigas*) or 'poc', which is known only from Lake Atitlán in Guatemala (Fig. 4.9). This bird is of particular biological interest because of its unusually strong bill and massive jaw musculature which seem to be an adaptation for feeding on freshwater crabs. The introduction of the largemouth bass (Fig. 4.8a) and its relative the smallmouth bass (*Micropterus dolomieui*) in 1958 and 1960 to promote a recreational fishery, coincided with a marked decline in the grebe populations, with numbers reaching a minimum in 1964. The bass appeared to compete with the grebes for crabs and fish, and were suspected also of attacking young birds. Fortunately, in this case energetic countermeasures against other threats to the birds such as uncontrolled reed cutting, coupled with an unexplained decline in bass populations, seem to have allowed a recovery of the grebe populations in more recent years (Polunin 1969, La Bastille 1974).

THE BLUE DUCK. In New Zealand, the blue duck (*Hymenolaimus malacorhynchus*) is another aquatic bird with specialised feeding habits which appears to have been affected adversely by introduction of exotic fish. The duck feeds on stream insects which it collects in clear streams by probing amongst stones. The bill has curious flexible flaps

a   b

Fig. 4.10. (a) The takahe or flightless rail (*Notornis mantelli*), (b) the kakapo or flightless parrot (*Strigops habroptilus*): two endemic species from New Zealand threatened by the grazing and browsing activities of introduced deer and chamois.

at the sides which are thought to serve a protective function (Kear & Burton 1971). Brown trout were introduced into New Zealand in the 1870s and feed on a range of invertebrate animals which is very similar to that consumed by the blue duck. Competition between the two species is therefore to be expected and it is probably no coincidence that the range of the blue duck has decreased at the same time as that of the introduced brown trout has expanded. The remaining strongholds of the duck are generally in mountain streams which have remained free from trout (Kear & Burton 1971).

*Effects of introduced mammals and birds
on the native fauna of New Zealand*

New Zealand is also notable for a large number of mammals which have been introduced for recreational hunting. These species include chamois (*Rupicapra rupicapra*) from Europe, tahr (*Hemitragus jemlahicus*) a goat-like species from the Himalayas, and at least seven species of deer from various parts of the world. Apart from two species of bats, New Zealand had no native mammals to suffer competition from the newcomers. However, the native vegetation both in forest areas and in the alpine regions has proved extremely vulnerable to grazing and browsing damage (Wodzicki 1961). In addition to accelerating soil erosion, this vegetation damage has had adverse consequences for

67

some of the unique flightless birds which are characteristic of New Zealand. One of these is the takahe (*Notornis mantelli*) (Fig. 4.10a) a flightless rail, which was once widely distributed in both North and South Islands, but now survives only in the mountains of the Lake District on South Island (Reid 1969). During most of the year the takahe feeds on alpine tussock grass (*Chionochloa*), but in winter moves into adjacent beech forest to take grasses, herbs and fern rhizomes. In both situations the birds are obliged to compete with introduced deer for their plant food and this represents a continuing threat to their survival (Mills 1976). Another species endemic to New Zealand, the flightless parrot or kakapo (*Strigops habroptilus*) (Fig. 4.10b) appears to be similarly threatened by competition for food from introduced deer and chamois.

Control of animal introductions

The disruption of natural communities by introduced species must rate as one of the most serious side effects of recreational hunting and fishing. Since many sportsmen have been unaware, or have chosen to ignore, the wider implications of their actions, the present trend for tighter control on animal introductions is to be welcomed (IUCN 1968, Nature Conservancy Council 1979).

Intensification of hunting caused by the tourist souvenir trade

A different source of disturbance to biologically interesting animal species comes from tourist demands for wildlife souvenirs. This can represent such a powerful stimulus for local hunters that survival of the exploited species is seriously threatened.

Historically a classic case is provided by the extinction of the huia (*Heteralocha acutirostris*) in New Zealand (Halliday 1978). This bird was of special interest to biologists because the bills of the two sexes were differently shaped and used in a complementary fashion in searching for wood-boring insects (Fig. 4.11). The white-tipped tail feathers on the otherwise black plumage were highly prized by the Maoris as a badge of high status and a mark of mourning. The hunting of birds for this purpose came under the jurisdiction of the local

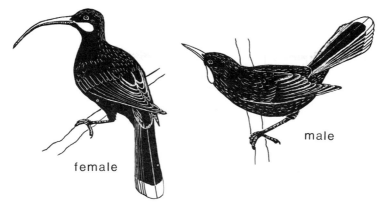

female

male

Fig. 4.11. The huia (*Heteralocha acutirostris*), an endemic species from New Zealand driven to extinction by the souvenir trade (after Halliday 1978).

'Tohunga' or priest and was carefully regulated. The situation changed, however, with the arrival of the Europeans. Their interest in items such as mounted huias as English drawing room exhibits and Maori headdresses made out of huia feathers persuaded the Maori hunters to abandon their former customs of careful husbandry to satisfy the new trade. In one instance eleven Maori hunters are reported to have collected no less than 646 huias in one month. Coupled with habitat changes and the toll taken by introduced predators, this commercial hunting activity was sufficient to drive the species to extinction.

A more recent example of the distortion of balanced relationships is provided by the trade in mounted animals in Taiwan. This has stimulated local hunters to exploit the fauna with an intensity far in excess of their original subsistence requirements. Taiwan's fauna is of special interest because it has diverged in isolation from that of the nearby mainland. It includes two species of pheasants, Swinhoe's pheasant (*Lophura swinhoei*) (Fig. 4.12) and the Mikado pheasant (*Syrmaticus mikado*) which occur nowhere else in the wild state. There is also on the island a distinct race of the sika deer (*Cervus nippon*). Wayre (1969) has described how mounted specimens of these and other endangered species have frequently been offered for sale at tourist resorts such as Sun Moon Lake in the centre of the island. 69

Fig. 4.12. Swinhoe's pheasant (*Lophura swinhoei*), a rare species from Taiwan which has suffered from the intensification of hunting associated with the tourist souvenir trade.

This souvenir trade, coupled with the effects of deforestation, obviously poses a serious threat to the survival of a number of important species.

Similar problems have arisen in the quite different arctic habitat of the Bering Sea area off the Western coast of Alaska. Here eskimo groups have a longstanding tradition of hunting walruses (*Odobenus rosmarus*) for meat and skins, and of using the tusks for traditional ivory carving. In recent years, the market for ivory items has caused some of these groups to kill walruses only for the tusks, and virtually to abandon the stable and culturally rich relationships that previously existed between hunters and walruses (Strickland 1981).

In the case of another marine animal, the hawksbill turtle (*Eretmo-chelys imbricata*) (Fig. 4.13), the threat comes from a combination of the local souvenir market for mounted immature specimens, and the world trade in 'tortoiseshell' (King 1982). Large numbers of adult and juvenile hawksbills are stuffed or freeze-dried in countries such as Indonesia, the Philippines, the Seychelles and Thailand (Mack *et al.* 1982).

It has so far proved extremely difficult to control tourist-induced activities of this kind or to cause them to be operated on a proper sustained-yield basis. In these circumstances the only solution is for

Fig. 4.13. The hawksbill turtle (*Eretmochelys imbricata*), a species threatened by the souvenir trade in 'tortoiseshell' and mounted juvenile specimens.

individual tourists to discourage the wildlife-souvenir trade by not participating in it.

Lead shot, lead weights and discarded fishing lines

The remaining group of adverse effects caused by hunting and fishing is connected with the field sportsman's equipment. A simple example is provided by the entanglement of waterbirds with lengths of nylon line discarded by fishermen. The line can become wrapped around the bird's feet, wings or bill and by interfering with normal activities cause death by starvation. A variant of the problem is where birds incorporate lengths of fishing line into the fabric of their nests. This can result in the entrapment of both adults and young, and in Britain has been identified as a cause of mortality amongst riverine birds such as the dipper (*Cinclus cinclus*) and coastal nesting species such as the gannet (*Sula bassana*). The extent of the discarded line problem has been demonstrated recently in a survey carried out by the Royal Society for the Protection of Birds. This showed that along a mile of river bank there could be as much as 800 feet (244 metres) of discarded line and that around some heavily fished lakes, the figure can exceed 3000 feet (914 metres).

Lead-poisoning in waterfowl

A related problem is that arising from the use of lead shot-gun pellets and lead fishing weights. Waterbirds are liable to ingest these items in the course of normal feeding or whilst they are searching for grit to take into the gizzard to aid digestion. In the acid and abrasive conditions of the bird's digestive tract, soluble lead salts are formed and produce symptoms of lead-poisoning. Although the severity of these symptoms is influenced to some extent by the type of food being eaten, it is determined mainly by the number of pellets ingested (Bellrose 1975). With one or two pellets in the gizzard, a bird may suffer only a temporary weight loss and recover subsequently. As the pellet load increases, however, fatal lead-poisoning becomes progressively more likely. The obvious symptoms are failure to feed, loss of weight and the production of bright green droopings, this latter effect being caused by excessive bile secretion. Swans in the later stages of lead-poisoning show a characteristic 'kinking' at the base of the neck which is caused by degeneration of the neck muscles (Fig. 4.14a).

Various attempts have been made to explore the extent of the problem. In Britain the examination of gizzards from more than 2000 wild ducks showed that over 10% of individuals of some species contained ingested shot-gun pellets (Mudge 1983). The list was headed by two diving species, the tufted duck (*Aythya fuligula*) (11.7%) and the pochard (*Aythya ferina*) (10.9%). A lower incidence (4.2%) of pellets was found in the mallard (*Anas platyrhynchos*) a species which feeds by 'dabbling' in more shallow waters. In North America a more extensive investigation of more than 36000 gizzards similarly showed an incidence exceeding 10% in four diving species and an incidence of between 5 and 10% in three dabbling species (Bellrose 1975). As a proportion of the birds with only one or even two pellets in the gizzard may ultimately survive, it is not easy to interpret these figures in terms of overall mortality. However, in the United States the Fish and Wildlife Service estimates that from 1.6 to 2.4 million ducks die every year as a result of ingesting lead pellets.

Lead-poisoning in swans

In Britain special concern has also been expressed about the effects of lead-poisoning on certain non-game species. In recent years there has been a noticeable decline in the populations of mute swans (*Cygnus olor*) in rivers flowing through urban and suburban areas. The famous swan herd on the River Avon at Stratford which numbered 80 in the winter of 1963, had by 1978 declined progressively to only four (Hardman & Cooper 1980). Similarly the swans on the River Thames, which have been counted annually since 1823 in 'swan upping'

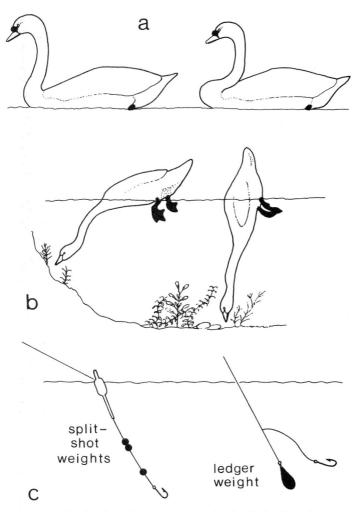

Fig. 4.14. Lead-poisoning hazards to swans associated with lead weights abandoned by fishermen. (a) Healthy swan (right) compared with individual showing 'kinked' neck characteristic of lead-poisoning (left). (b) Bottom-feeding habits of swans which make them liable to ingest lead items. (c) Diagram illustrating some typical uses of angling weights.

73

ceremonies, declined during the 1970s from a mean level of between 400 and 600, to only 153 in 1981. There is little question that lead poisoning is a major factor in these and other situations where dramatic declines have taken place (Nature Conservancy Council 1981).

In rivers close to centres of population the injurious lead items included both the split-shot weights which anglers pinch onto their lines and also the larger 'ledger' weights (Fig. 4.14c). Large quantities of split-shot enter rivers as a result of being spilled from shot dispensers, and weights of all kinds are discarded on abandoned lines. Swans feeding on the river bed (Fig. 4.14b) readily ingest these items which they mistake for the grit normally taken in for digestive purposes. A single ledger weight is more than sufficient to cause the death of a swan by lead-poisoning.

The importance of fishermen's weights in causing swan mortality can be gauged from the fact that of the 70 dead swans examined in recent years from the River Avon 77 % were judged to have died from this cause (Nature Conservancy Council 1981). A similar examination of 118 swans dying on the River Trent attributed 90 % of this mortality to poisoning from fishermen's weights (Simpson *et al.* 1979). In areas where both duck shooting and angling occur, swans, which otherwise enjoy legal protection, are in jeopardy both from fishermen's weights and from spent shot-gun pellets. Even in parts of East Anglia, where substantial no-shooting refuges are available, swans ingest pellets and weights in sufficient quantities to account for 29 % of their total mortality (Owen & Cadbury 1975). In this area, not only are mute swans affected, but also important overwintering populations of Bewick's and whooper swans (*Cygnus columbianus* and *Cygnus cygnus*).

Waterbird poisoning from these causes is regrettable on many counts, the most important being that it condemns many millions of birds to a lingering death and puts further pressure on species already threatened by other factors.

Remedial measures

There are in fact no significant technical difficulties involved in producing angling weights from non-toxic substances such as tungsten polymers and steel-based putties, and in Britain increasing use is being made of these materials. The situation with shot-gun pellets is more complicated. Steel pellets, because of their lower density are less effective in bringing down birds than lead ones although the difference is only marginal (Mikula *et al.* 1977). They can, however, cause accelerated wear to the barrels of certain types of guns and this is an aspect of the problem which needs further investigation. Nonetheless the use of steel shot has already been made mandatory in certain National Wildlife Refuges in the United States, and this type of measure should be considered for wider application.

Recreational hunting in perspective

If the same criteria are applied to recreational hunting as to other recreational pursuits, namely that they should be practised with minimal environmental damage and interference with other interests, then the record for hunting is generally a poor one. The continued failure to take a responsible attitude to introductions of game species and to unjustified 'control' of predators, are major grounds for criticism, as also is the tardiness with which the lead-contamination problem has being tackled. On the other hand there is validity in the sportsman's claim that, equipped with suitable biological data, it is possible to exploit game-animal populations without destabilising them. It can further be argued that recreational hunters are able to give useful assistance to other land users in helping to control species that have become locally so numerous as to cause crop damage. In the United States, forest managers in many areas produce maps to indicate where damage to trees by deer is severe and therefore where sportsmen might usefully concentrate their attention.

Sometimes, however, this latter rationalisation of hunting as a means of pest control is a spurious one. In his historical review Thomas (1983) has pointed out how in Britain, fox-hunting is frequently represented as a pest-control measure, in spite of the fact that there is a long history of preserving fox cubs, importing foxes from adjacent counties and planting vegetation for the animals' benefit.

A consideration of the relationship between ethics and hunting falls outside the scope of an ecological discussion, but it is well known that there are opposing schools of thought. Whilst admitting that modern man no longer needs to hunt animals for food, the supporters of recreational hunting argue nevertheless that opportunities are required in the modern world for man to practice and enjoy some of the hunting and tracking skills used by his forbears. The opposing view is that hunting for pleasure is unacceptable on moral grounds. This position has gained force from the well formulated case advanced by Singer (1975). He argues that recreational hunting is an aspect of 'speciesism', a widely assumed but unwarranted human attitude that man can deal with other species in any way he finds expedient. Obviously the future of recreational hunting will be influenced as much by the outcome of this debate as by ecological considerations.

5.
ENJOYMENT
OF
SCENERY

Although the enjoyment of scenery is ostensibly a more passive recreational pursuit than some of the others already discussed, it too can represent a source of environmental disturbance. It has become apparent for example, that large numbers of visitors congregating in scenic areas can cause extensive and unsightly damage to soils and vegetation (Willard & Marr 1970, Burden & Randerson 1972, Liddle 1975). Moreover the facilities established to serve these visitors may further detract from scenic quality because of their poor design or inappropriate siting (Zaslowsky 1983). For some visitors even the presence of other people apparently can interfere with enjoyment, although this is a factor which is difficult to evaluate in a satisfactory way (Stankey 1978).

An ecological approach can help to alleviate some of these problems by providing information about the attributes of various vegetation types and the conditions necessary for their growth. A simple example of this type of input is provided by the design of vegetation screens to camouflage potential recreational eyesores such as car parks and caravan sites. When designing these planting programmes it is clearly important to match the choice of tree and shrub species to the local soil, to take account of specific factors like salt-spray in coastal areas which limit the choice of species and to relate growth-form to screening requirements (Ministry of Housing and Local Government 1962).

A more general application of ecology involves the assessment of damage to vegetation and soil caused by trampling, and the design of measures to combat this form of disturbance.

Trampling damage to vegetation and soils in scenic areas

The susceptibility of different plant communities

Trampling damage is frequently unwelcome in scenic areas because it destroys attractive plant communities and creates eroded, unsightly soil surfaces. In temperate regions where most of the work on trampling effects has been carried out, it has become apparent that not all plant communities are equally vulnerable. Studies in the Rocky Mountains of North America, for example, have shown that with a given intensity of use the damage to trailside vegetation is greater in forests than in grasslands (Weaver & Dale 1978, Cole 1978) (Fig. 5.1). A probable explanation for this difference lies in the fact that many forest-floor plants have large leaves and thin cell walls to allow them to utilise the lower light intensities within the forest and consequently are particularly vulnerable to mechanical damage. By contrast the species which characterise open grassy habitats often have attributes which increase their tolerance to trampling; these include flat rosettes of leaves, or in the case of grasses, growing points which are hidden below soil level during the winter (Bates 1935) (Fig. 5.2).

At elevations above the tree-line, alpine flowers are an important attraction for visitors. Unfortunately, however, many of these plants prove to be particularly vulnerable to damage. In the Rocky Mountain National Park (highest elevation 3798 m), where the construction of a high level road allowed the public easy access to alpine meadows for the first time, as much as 95% of the vegetation cover has been destroyed in areas close to the road. The only plants able to survive at these sites have been those growing in the vicinity of protective boulders. Even in less heavily used areas, trampling damage has been sufficient to flatten cushion plants and to prevent many species from flowering (Willard & Marr 1970, 1971). Amongst the reasons suggested for the susceptibility of alpine vegetation to damage are the slow rates of growth and recovery associated with low environmental temperatures, and the slow rate at which new soil develops on damaged areas (Billings 1973).

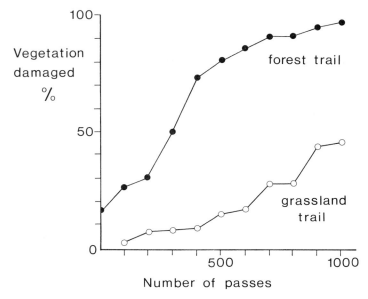

Fig. 5.1. Results of a study from the Rocky Mountains showing how, for a given level of trampling, vegetation damage is greater on forest than on grassland trails. (Adapted with permission from 'Trampling effects of hikers, motor cycles and horses in meadows and forests', T. Weaver & D. Dale, *Journal of Applied Ecology*, **15**, 451–7. Blackwell Scientific Publications.)

Fig. 5.2. Great plantain (*Plantago major*) growing along a path edge in a Cardiff City park. The tough texture and rosette arrangement of leaves make this species resistant to trampling damage.

79

Other studies have confirmed the vulnerability of the more showy alpine flowers. They have revealed, however, that some of the less conspicuous plants such as grasses and sedges can be surprisingly resistant to damage. In North America, studies in the Olympic National Park in Washington State have demonstrated that the sedge *Carex nigricans* comes into this category (Bell & Bliss 1973). In the Austrian Tyrol similar attributes have been found in another sedge species, *Carex curvula* (Grabherr 1982).

Management implications

Although it is possible in this way to analyse the relationship between trampling pressure and vegetation change, account has to be taken of other factors when translating such information into management terms. A site manager may decide for example that although routing visitors along forest trails rather than grassland ones is likely to cause more vegetation damage, this is justified because the visitors are less conspicuous to distant observers.

The term 'carrying capacity' has been widely employed to denote a level of visitor use beyond which environmental deterioration is likely to occur. However, as Burden & Randerson (1972) have pointed out, this is as much a managerial concept as an ecological one. Any one of a number of carrying capacity levels can be applied to a site according to the management objectives selected. Thus if the aim is to retain visually attractive but sensitive species in a sward, trampling needs to be kept to a minimum. If, on the other hand, it is judged sufficient to maintain a fairly continuous vegetation cover, irrespective of species composition, then a higher level of use becomes permissible. In practice there can be a tenfold difference between the acceptable carrying capacities in the two cases.

In situations where no obligation is felt to restrict the composition of the sward to indigenous species, the carrying capacity can sometimes be increased further by applications of fertiliser and the sowing of commercial grass-seed mixtures. Experience derived from the management of sports turfs and the lawns of urban parks indicates that useful species for this purpose include perennial rye grass (*Lolium perenne*), annual meadow grass (*Poa annua*), smooth-stalked meadow grass (*Poa pratensis*) and timothy grass, (*Phleum pratense*) (Canaway 1980).

Fig. 5.3. (a) Soil erosion and exposure of tree roots caused by trampling at Friar's Crag, a vantage point on the shore of Derwentwater in the Lake District National Park. (b) Remedial measures involving the fencing of restored areas and the routing of visitors along paths surfaced with wood chips.

For any trail there is a level of visitor-use beyond which normal soil and vegetation cover can no longer be maintained. It may then become appropriate either to divert visitors from sensitive areas or to create an artificial surface of some kind. A good example of a diversion strategy is provided by the provisions made at Tarn Hows, 81 a much-visited beauty spot in the Lake District National Park in northern England. Here by realigning the footpaths and altering the pattern of car parking it has been possible to direct visitors away from a readily eroded steep bank overlooking the lake (Brotherton *et al.* 1977).

If the decision is taken to provide an artificial surface, a wide range of techniques is available. In the alpine meadows of the Rocky Mountains National Park, a system of asphalt-surfaced paths has been devised to allow visitors to see the flowers without causing further damage. In Britain, in the North York Moors National Park the erosion of paths on deep peat soils has been checked by laying limestone chippings on plastic sheeting. In the Lake District National Park lakeside paths on the shores of Derwentwater have been surfaced with wood chips to avoid further damage to the roots of the pine trees (*Pinus sylvestris*) which are a special feature of the site (Fig. 5.3). In many mountain areas deep gullies caused by path erosion have been refilled and surfaced using bedded stones or rock chippings. In these circumstances it is frequently necessary to construct underpath culverts to protect the rehabilitated path sections from water damage.

Site managers have to face the possibility that not all visitors will regard artificially surfaced paths as a visual improvement on worn paths. In some mountain areas of Britain there has been recent criticism of the use of intrusive brick and pre-cast concrete materials in path construction (Hutchinson 1985). Ecology can be of little assistance in resolving this kind of dilemma in which aesthetic judgements and decisions about the practicality of various constructional techniques are prime considerations.

Environmental changes at historical and archaeological sites

Similar issues have arisen at archaeological and historical sites. At Stonehenge, Britain's most-visited archaeological site, increasing use by visitors had by the early 1960s substantially destroyed the grass cover within the circle of stones and on the approach path. The reaction of the custodians of the site (then the Ministry of Works) was to remove the remaining grass and topsoil and substitute a covering consisting of furnace clinker with a top layer of gravel

Fig. 5.4. Viewing arrangements at Stonehenge where the risks of damage to the monument and its immediate surroundings are minimised by directing visitors to a tarmacadam-surfaced area (left) and a perimeter path (foreground).

(Chippindale 1983). Although this certainly provided a durable surface it soon became apparent that it was not an appropriate solution, because gravel fragments embedded in visitors' shoes were scratching the recumbent stones. As part of the archaeological interest of the monument lies in the presence of surface carvings, this type of damage was clearly unacceptable. In 1978 the decision was therefore taken to restore the turf at the centre of the circle and to prevent any regular close approach by visitors. The monument can now be viewed from a tarmacadam path approaching in a loop from the north west, or from a more distant perimeter path (Fig. 5.4). The latter, which is unsurfaced, periodically becomes sufficiently worn to require realignment.

Whilst such arrangements undoubtedly have the desired effect of protecting the monument and its immediate surroundings from damage, few would feel that they represent an imaginative approach to displaying and interpreting the archaeological interest of the site. A study group which was recently established to examine the issue concluded that Stonehenge should be interpreted in relation to other archaeological features in the vicinity and that a network of paths

83

should be devised to allow visitors to move freely between the various points of interest (English Heritage 1985). These proposals recognised that heavily used path sections might need to be surfaced unobtrusively or periodically realigned.

A more controversial scheme currently being examined is to divert visitors from the real Stonehenge to a plasticised facsimile, predictably dubbed 'Foamhenge' by the press. Whether or not this represents an appropriate option at Stonehenge, it is an approach which has been adopted successfully at other archaeological sites and most notably at the painted caves at Lascaux in Southern France. The problems associated with visitor pressure at this latter site have also involved changes in plant communities, but of a rather unusual kind.

Lascaux

Caves in general are vulnerable to environmental changes caused by the pressure of visitors. The passage of people through small chambers and narrow passages is known to produce localised increases in temperature and carbon dioxide levels (Merenne-Schoumaker 1975). If, in addition, artificial lights are installed to illuminate cave features, favourable conditions are created for the growth of green plants such as mosses and algae. These would normally be absent from the cave environment and their growth can create unsightly brown and green mats on otherwise gleaming white calcite structures such as stalactites, stalagmites and flowstone curtains (Fig. 5.5). In one study it was shown that of the 200 lights present in a cave system, 77.5% had algae and 20.5% had moss growths associated with them (Stark 1969).

The presence of plant growths has especially serious implications in caves such as those in France and Spain which are adorned with prehistoric wall paintings. The paintings are remarkable from an aesthetic standpoint and also because they portray animals which occupied these regions during the Ice Ages, but have long since become extinct (Laming 1959, Sieveking & Sieveking 1962).

At Lascaux the first indications that tourism was altering the cave environment came from measurable increases in carbon dioxide levels in the chamber and consequent complaints of dizziness amongst visitors. The reaction of the management was to install air-conditioning

Fig. 5.5 Algal growth in the vicinity of floodlights in the show cave at Dan yr Ogof in South Wales. Such growth seriously disfigures white calcite structures.

equipment to minimise this source of discomfort. At this stage little thought was given to the possible biological effects of the floodlight system used to illuminate the paintings. In September 1960 there were the first intimations of a more serious crisis when a small patch of green was noticed on one of the paintings. In the weeks that followed other patches appeared and began to spread. In spite of the fact that the authorities reacted by closing the caves, the affected area continued to expand until it covered a substantial proportion of many of the important paintings (Fig. 5.6). On investigation the growths proved to be dominated by a small sphaerical alga of the genus *Palmellococcus* (Leferve & Laport 1969). It was found that the organism could be controlled by spraying the walls with dilute formaldehyde solution. However, the pungent odours associated with this treatment ruled out the simultaneous use of the cave by tourists.

In 1963 the authorities reluctantly reached the conclusion that large-scale tourist viewing of the paintings was in fact incompatible with their continued preservation, and the decision was taken to close the caves to tourist entry. Subsequently a facsimile of the Lascaux 85

Fig. 5.6. Disfigurement of wall paintings by algal growth (shaded) in the 'Hall of Bulls' at Lascaux, showing the maximum extent of the affected area before remedial measures were taken. The animals portrayed include deer, wild cattle and horses, and an imaginary animal (far left) (after Lefevre & Laport 1969).

Cave, complete with accurately contoured concrete walls and reproductions of all the paintings, has been erected in a nearby quarry at the cost of £500000 and is proving a major tourist attraction.

Perpetuating plant formations of scenic importance

The other major contribution of ecology to scenic amenity lies in the development of management regimes for perpetuating plant formations of high scenic value. In this connection the ecological concept of 'plant succession' is of particular relevance.

Plant succession

The mixture of plant species which first colonises a newly exposed area of bare soil usually differs considerably from the plant community which develops eventually on the site. In wetter climates and in the absence of human intervention the initial colonising flora of herbaceous plants is replaced first by shrubs and subsequently by forest trees (Fig. 5.7). The final stage of the succession, in this case a forest, is referred to as the climax community. In regions where the annual rainfall is limited, grasses dominate climax communities, and trees may be sparse or absent. In their natural form the prairies of North America and the steppes of Eurasia are examples of vegetation types which fall into this category. Land managers sometimes fail to perceive the distinction between open vegetation types which represent climax communities, and those which are open because they are being held at an intermediate successional stage by the action of grazing stock (Fig. 5.7).

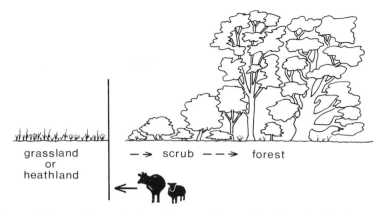

grassland   — →   scrub — — →   forest
or
heathland

Fig. 5.7. Diagram illustrating how the natural succession of grassland and heathland to scrub and forest can be checked by the action of grazing stock.

Management of grassy parkland

A striking demonstration of the role of grazing animals in preventing the encroachment of grasslands by shrubs and trees, is provided by the changes recorded during the last fifty years in parkland areas adjacent to the cities of Berkeley and Oakland in California.

When the parks were designated in 1934 a policy was adopted of excluding grazing livestock. The result has been a progressive invasion of the open grassland by shrubs and scattered trees as the vegetation develops towards its natural climax type (Fig. 5.8). The principal invasive shrub has been the species variously known as coyote brush or chaparral broom (*Baccharis pilularis*). By resurveying areas for which early maps and photographs are available it has been possible to show that there has been a three- to five-fold increase in shrub cover over large areas of the park system (McBride & Heady 1968, McBride 1974). These changes are widely regarded as reducing the recreational value of the parks because the bushes impede the free movement of hikers and horse riders, and obscure the view. An added complication arises from the fact that the shrubs create a favourable habitat for the development of poison oak (*Rhus diversiloba*) (Fig. 5.9), a low woody plant which can cause serious allergic reactions if it comes into contact with the skin (Raven 1966, Tampion 1977). To return these areas to grasslands will require a massive programme of cutting

87

Fig. 5.8. Progressive invasion by coyote brush (*Baccharis pilularis*) after withdrawal of grazing stock from the Redwood Regional Park near Oakland, California. Photographs show the situation 6 years (a) and 33 years (b) after stock removal (photographs by courtesy of Professor J. McBride, University of California, Berkeley).

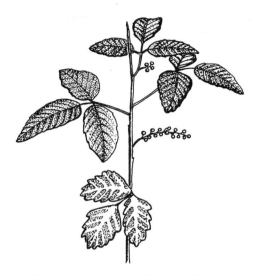

Fig. 5.9. Poison oak (*Rhus diversiloba*) (after Raven 1966).

and burning to remove shrub growth and the reintroduction of stock-grazing.

In Britain a somewhat similar situation arises in a number of Country Parks. These are designed principally to serve the recreational needs of visitors from urban areas. They are frequently designed to include large expanses of short turf suitable for picnicking and informal recreation by family groups. Without active management such grassy areas become tussocky and unpleasant to use and are later invaded by scrub.

Although one option is to check these successional processes using mechanical cutting techniques, there is an increasing tendency to favour the use of grazing animals (Lowday & Wells 1977, Wells 1980). Sheep, cattle, ponies and deer have all been used for this purpose and Table 5.1 lists some of the advantages and disadvantages of the various animal species. In a number of the Country Parks derived from private estates, semi-tame herds of park deer are already present. In such cases the obvious strategy is to continue using these animals for vegetation management (Fig. 5.10).

89

Table 5.1. *Factors relevant to the choice of animal species for vegetation management in recreational areas (based on Lowday & Wells 1977)*

	Sheep	Cattle	Horses and ponies	Deer
Risk of injury to visitors	Low	Low, possible alarm from inquisitive behaviour	Appreciable risk from kicking and biting	Low, possible aggression from rogue stags
Susceptibility to worrying from visitors' dogs	High	Low, although calves vulnerable	Low	Low
Other drawbacks	Disease prone on poorly drained areas	Cow pats offensive. Can poach grass in wet weather	Selective grazer, may produce patchy appearance in vegetation	Overstocking may reduce floristic diversity of vegetation
Management inputs required	High especially at lambing time	High at calving time. High if dairy herd	Low	Additional fence maintenance, need for controlled culling regime
Additional benefits from saleable products	Revenue from wool and meat	Revenue from hides meat and sale of live animals	Revenue from riding hire and sale of live animals	Revenue from hides, antlers and venison
Situations of special suitability	Irregular terrain with steep slopes	Level terrain where control of coarse grass required	Enclosed paddock with public contact only at fence	Established parkland with existing high fencing

Fig. 5.10. The use of fallow deer (*Dama dama*) for vegetatation management in Margam Country Park in South Wales. Grazing by deer prevents the development of grass tussocks and scrub in recreational zones such as the children's play area shown in the lower photograph. Young trees need to be protected from browsing and fraying damage by the deer (photographs by courtesy of Margam Country Park, West Glamorgan County Council).

Management of heather moorland and flower-rich meadows

Grazing animals play an important part in maintaining other types of scenically important vegetation. In Europe heather moorland provides a notable example of this kind. In autumn the large expanses of purple flowers of the dominant heather species (*Calluna vulgaris*) frequently provide a notable landscape feature. The relationship between moorland vegetation and grazing pressure has been investigated extensively and it is now well established that heather moorland can be perpetuated only under moderate grazing regimes (Gimingham 1972, Institute of Terrestrial Ecology 1978). Heavy grazing pressure is likely to result in the heather being replaced by rough grasslands, whereas reduction or discontinuation of grazing usually leads to encroachment by shrub and woodland (Froment 1981, Ball *et al.* 1982). In scenic areas either outcome is likely to be regarded as less desirable than the original heather cover.

Similar considerations of optimum grazing levels arise in relation to the management of chalk and limestone grasslands in Western Europe. Here the aesthetic attraction is based on a range of flowering plants rather than on a single species. In an experiment on chalk grassland in the Netherlands in which grazing animals were excluded from a plot for 10 years, it was found that by the end of this period, hawthorn shrubs (*Crataegus monogyna*) and bird cherry trees (*Prunus avium*) had invaded the area. Moreover the numbers of visually interesting flowering-plant species had also declined (Willems 1983). Comparable changes have been observed on chalk grasslands in Britain where grazing has ceased. Ungrazed pastures are first invaded by scrub species such as hawthorn, dogwood (*Cornus sanguinea*) and blackthorn (*Prunus spinosa*), and subsequently by forest climax vegetation dominated by beech (*Fagus sylvatica*) or pedunculate oak (*Quercus robur*) (Wells 1969, Duffey *et al.* 1974).

In these situations, maintaining the vegetation at the desired successional stage requires the careful control of grazing levels. If this cannot be achieved within the framework of existing agricultural enterprises, it then becomes necessary to manage flocks of grazing animals principally for amenity purposes or to embark on a labour-intensive programme of mechanical cutting (Wells 1970).

Although the majority of open recreational habitats in temperate

Fig. 5.11. Sheep overwintering in a Welsh oak wood.

regions require this kind of treatment to prevent scrub encroachment some exceptions need to be noted. Scrub and tree growth is not a problem in mountain habitats above the natural tree line, because environmental conditions are too severe. The short-grass prairies of North America, the short-grass steppes of Eurasia and the 'grassveld' regions of South Africa also represent habitats which are generally free from scrub encroachment. In these situations low rainfall and soil dryness are the principal inhibiting factors, although the occurrence of natural fires may be important in some circumstances (Eyre 1963).

The management of woodland vegetation of scenic interest

Whatever the value of domestic stock in maintaining open habitats, grazing animals have to be regarded in a different light when considering management regimes for woodland vegetation. In Britain's National Parks where fragments of once-extensive oak woodlands contribute significantly to the landscape, sheep grazing is now regarded as a serious problem. In upland regions, oak woods are widely used to provide grazing and shelter for sheep in the winter (Fig. 5.11). Unfortunately, this practice causes developing tree

93

seedlings to be bitten away by the animals and allows very few to survive to the sapling stage. In one survey of relict woods in the Snowdonia National Park in North Wales, it was found that tree saplings were absent from 80% of wooded areas (Smith 1982). In the absence of remedial actions these heavily grazed woods will disappear at the end of the life span of the existing trees.

It is now accepted that the appropriate solution is to construct a series of sheep-proof enclosures in the areas most severely affected (Piggot 1983). This allows regeneration to take place without completely denying the farmer the use of the woods for stock-grazing.

Other causes of regeneration failure in tree stands

Stock-grazing is by no means the only reason why some tree stands fail to regenerate themselves. With many species, seedling development is inhibited by the shade of existing trees, so that new growth can take place only where old trees have fallen or have been removed, thus allowing more light to reach the forest floor (Shaw 1974). In a few instances even these conditions are not sufficient to guarantee regeneration, because tree seedlings are suppressed by competition from aggressive ground-cover species (Connell & Slatyer 1977).

Again, park management studies from the San Francisco region of California provide some interesting case histories (McBride & Froehlich 1984). In the city parks of San Francisco there are some striking stands of Monterey cypress (*Cupressus macrocarpa*) and Monterey pine (*Pinus radiata*). These were planted 70 – 100 years ago, and being relatively short-lived trees are now approaching the end of their natural life spans. Unfortunately, in many cases, because the trees are of similar age and closely grouped, natural regeneration has been prevented, and the park authorities have come to realise that it may now be impossible to maintain a continuous presence of mature trees on these sites. As the existing trees die or have to be removed because they are unsafe, there is likely to be an appreciable time interval, possibly as long as 50 years, before the seedlings now being established will have grown into mature trees. With hindsight the appropriate management regime (which should have been started from the outset) would have been continually to remove trees on a rotational basis, to allow new growth to develop in the gaps. This would have guaranteed a continuous presence of at least a proportion of mature trees. It is important not to underestimate the significance attached by the public to mature trees in scenic areas. Such trees frequently have more interesting and varied canopy shapes and more interesting bark patterns than partially grown specimens.

The San Francisco parks are also instructive in illustrating the impediment to tree regeneration which can be caused by the presence of aggressive ground-cover plants. In a number of the parks, even where otherwise suitable canopy openings exist, tree regeneration has apparently been suppressed by competition from ground-cover species such as German ivy (*Senecio mikanioides*) and Algerian ivy (*Hedera canariensis*) (McBride & Froehlich 1984). Such vegetation needs to be cleared if tree seedlings are to have any chance of survival.

If one contribution of ecology to the management of scenic areas is considered to lie in the investigation of environmental changes caused by visitor use, these latter examples of vegetation management illustrate the second type of input. By interpreting traditional farm and forestry practices in the light of the ecological processes of plant succession and population recruitment, it becomes possible to point to ways in which these regimes can be adopted or simulated for vegetation management in scenic areas.

6.
DISEASE
HAZARDS

The growth of international tourism has brought with it a spectacular increase in the incidence of exotic diseases amongst travellers. For example, during the period 1975–80 at least 735 tourists from Britain are known to have contracted malaria in the course of their travels (Bruce-Chwatt 1982). Similarly, concern is being expressed about the rising incidence of schistosomiasis, amoebiasis and trypanosomiasis amongst visitors to the tropics (Domart *et al.* 1969, Woodruff 1975, Janssens & De Muynck 1977, Bruce-Chwatt 1978).

Even less-exotic tourist destinations generate their quota of illnesses, most notably the various forms of gastrointestinal disturbance referred to as 'travellers diarrhoea'. This is a problem encountered particularly by travellers from Western countries who visit areas where insufficient attention is given to food hygiene and the protection of water supplies (Geddes 1982, Mackay 1982). Contrary to earlier views that travellers diarrhoea was frequently a consequence of changes in diet, it is now increasingly apparent that infective agents are usually involved (Christie 1980). Some indication of the scale of the problem is indicated by a recent survey of subsequent illness amongst travellers returning to Scotland from package holidays taken abroad, mainly in the Mediterranean region. Out of 1961 people questioned, 766 (39%) had suffered some kind of gastrointestinal disturbance (Reid 1982).

Even people in temperate regions using recreational facilities close to their homes do not necessarily escape disease problems. After much argument it is now firmly established that bathing in sewage-polluted waters, whether at the coast or inland, can carry a significant health risk. Similarly both in Central Europe and in parts of North America it is becoming increasingly apparent that in some areas, visitors to woodland–edge trails and picnic sites are vulnerable to diseases transmitted by biting ticks.

The ecologists' contribution in this field is to help in identifying the environmental situations where disease transmission is likely to occur and, where possible, to suggest remedial measures. This kind of input is particularly relevant where animal transmitters (vectors) or animal reservoirs are implicated, or where 'indicator' organisms are being used to measure health risks.

Tropical disease

Tourists from temperate countries are likely to be at a double disadvantage when visiting the tropics. In the first place they usually lack the partial natural immunity to local diseases enjoyed by residents. Secondly, in unfamiliar surroundings they fail to appreciate which activities are likely to make them vulnerable to infection. A striking example of this second type of problem is provided by the risk of contracting schistosomiasis from inland bathing sites.

The schistosomiasis problem at freshwater bathing sites

The routine assumption made in temperate regions that rivers and pools which are shallow and clear are safe to use for bathing, cannot be applied generally in the tropics. This is because of the parasitic infection, schistosomiasis, which is widespread in Africa and the Middle East and also in certain parts of Asia, South America and the Caribbean Islands (Fig. 6.1).

The disease is caused by a parasitic worm which grows to maturity in man, but passes the early part of its life-cycle in a freshwater snail (Fig. 6.2). Eggs shed by the adult worm are discharged from the human host in faeces or urine. If they enter water, they hatch to produce a free-swimming larva (the miracidium) which enters the body of the snail. After further multiplication inside the snail, the parasite emerges in the form of another free-swimming larva (the cercaria). It is this stage which reinfects man by penetrating the skin. As a single 97

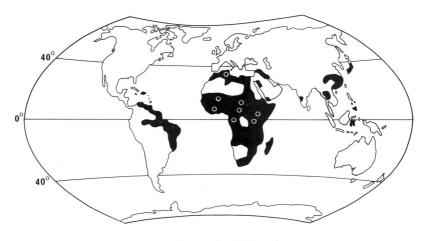

● sites of tourist infections

Fig. 6.1. Risk areas for schistosomiasis showing sites of recent infections amongst tourists.

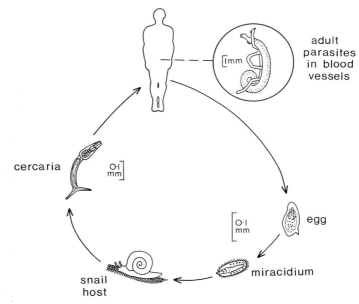

Fig. 6.2. The life-cycle of the parasite *Schistosoma*, responsible for schistosomiasis in man.

snail can emit 2000 cercariae daily and continue to do so for up to 200 days, the risks of bathing in waters containing infected snails are very considerable. Left untreated, schistosomiasis infections can have serious medical consequences. These arise mainly from surplus parasite eggs which circulate in the bloodstream and become lodged and encapsulated in various parts of the body such as the bladder, lungs or nervous system where they interfere with normal functions.

There are now many recorded instances of Western tourists contracting schistosomiasis apparently through ignorance of the problem. In a typical example from Western Sudan (Saeed & Magzoub 1974) thirteen Europeans contracted schistosomiasis after swimming in pools in the Jebel Marra mountains. Although these pools were ostensibly clean and aesthetically most attractive, they were also used by local children with schistosome infections and supported a large population of infected snails. In Ethiopia at least fifty cases of schistosomiasis amongst tourists have been associated with visits to the Omo National Park. Careful investigations have identified bathing pools at a river camp site as the most likely source of infection (Fuller *et al.* 1979, Zuidema 1981). The river supplying the pools was found to contain large numbers of schistosome-carrying snails, and it was concluded that people from nearby villages and locally recruited Park staff formed part of the transmission cycle. In Tunisia, one of the more unusual attractions offered to tourists is a camel ride across the desert to an oasis. At least twenty Italian and French tourists who took the opportunity to visit the oasis of El Mamoun in the south of the country were subsequently shown to have contracted schisto-somiasis as a result of swimming in the oasis waters (Domart *et al.* 1969, De Carneri & Bianchi 1970, Grassi 1970). Elsewhere in Africa, in Uganda, Mali, the Ivory Coast and the Central African Republic, there have been similar reports of tourists becoming infected when using local bathing waters (Domart *et al.* 1969).

In schistosomiasis areas the guiding principle which needs to be more widely publicised is that bathing should be restricted to chlorinated swimming pools, unless some authoritative assurance can be given that a natural site is safe. For example, some waters are safe because conditions are too acid to support snail populations. The fact that local people are using a site is no guarantee of its safety, indeed the reverse is more likely to be true. The schistosomiasis problem not only provides a striking example of the tourists' lack of awareness of an important health risk, it also illustrates the failure of tour operators and the managers of recreational facilities to draw attention to such hazards.

Mosquito-borne diseases in tropical cities

In view of the modern appearance of many tropical cities, the tourist might be excused for believing that the health hazards likely to be encountered would be little different from those associated with cities at home. Whilst this might be true for visitors who rarely stray from air-conditioned coaches and hotel rooms, it is less likely to be the case for those who understandably wish to gain a more first-hand acquaintance with the local environment, by visiting open-air markets and restaurants, or walking in local parks.

Even modern tropical cities still provide numerous breeding places for disease-carrying mosquitoes. The species which breed in organically polluted water find no shortage of this in open street drains and latrine sumps. Equally those mosquitoes which make particular use of small amounts of clean water frequently find this habitat in abundance amongst the water-storage containers used by local people, and in the rainwater that accumulates in tin cans, car tyres and other items of discarded rubbish. Figure 6.3 illustrates some of the typical mosquito species of tropical cities and the circumstances in which they are likely to be encountered.

FILARIASIS. *Culex pipiens fatigans* (also known as *C. pipiens quin-quefasciatus*) is the principal mosquito associated in its larval stage with the organically polluted water of open street drains, soakways and latrine sumps (Fig. 6.3d) (Singh 1967, Mattingly 1969). The adult mosquitoes bite at night and are responsible for transmitting the larvae of the parasitic worm *Wuchereria bancrofti* which causes filariasis in man. In extreme cases of this infection the worms block the lymphatic vessels and produce conspicuous swelling of the parts of the body affected, a condition known as elephantiasis.

MALARIA. The species which carry malaria are generally associated with clean-water breeding sites. In the older parts of cities in India, the malaria vector *Anopheles stephensi* breeds in water-storage tanks and cisterns. Its counterpart in Africa, *Anopheles gambiae*, has a

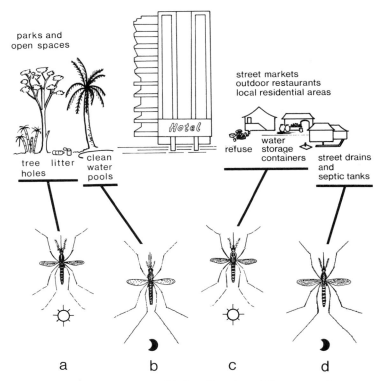

Fig. 6.3. Diagram to illustrate possible breeding sites of disease-carrying mosquitoes in tropical cities: (a) *Aedes albopictus*, vector for dengue in South East Asia (day biter); (b) *Anopheles gambiae*, vector for malaria and filariasis in Africa (night biter); (c) *Aedes aegypti*, vector for dengue in Africa, South East Asia and the Caribbean (day biter); (d) *Culex pipiens fatigans*, vector for filariasis in Africa, South East Asia and the Caribbean (night biter).

preference for open, sunlit water bodies which brings it into parks and similar open areas in cities (Fig. 6.3b). These malaria vectors are also night biters.

DENGUE FEVER. The most important day-biting mosquitoes are the various species of *Aedes* which are responsible for the transmission of dengue. This is a virus disease which occurs in a particularly severe form in South East Asia. The favoured larval sites for *Aedes aegypti* are the small accumulations of water which collect in discarded tin cans and car tyres. The mosquito also invades the water-storage jars used in houses without direct water supplies (Fig. 6.3c) (Rao *et al.* 1973). A survey of *Aedes aegypti* larvae in Singapore showed that waterfilled ant traps, earthenware jars, bowls, tanks, tin cans and drums were favoured habitats (Chan *et al.* 1971). Similar results were obtained in surveys by the *Aedes* Research Unit in Bangkok, Thailand. In South East Asia, the other dengue vector *Aedes albopictus* favours less-domestic situations, but still ones which are relevant to the tourist. It is prevalent in public parks and open spaces where the larvae occupy habitats such as waterfilled tree holes, cut bamboo stumps and discarded drink containers (Fig. 6.3a). Although *Aedes* mosquitoes are best known for their role in the transmission of yellow fever in Africa and South America, international vaccination requirements have virtually eliminated this disease as a hazard to tourists. On the other hand, dengue has increased in importance in parts of the Caribbean and Africa as well as in South East Asia, and merits more serious attention than it has received to date. Although short-term visitors appear to escape the severe haemorrhagic form of dengue which affects residents, the milder form is currently not uncommon amongst tourists returning from South East Asia.

Protective measures

Many factors are likely to influence the tourist's chances of contracting disease from mosquito bites. The effectiveness of the mosquito control measures carried out by the urban authority obviously has a direct bearing on the issue. So also does the number of infective bites necessary to establish an infection for each parasite. The vector of filariasis, *C.p. fatigans*, is difficult to control because of the abundance of polluted-water habitats and the development of resistance to insecticides in this species. However, these problems are partially offset by the fact that, with filariasis, a succession of infective bites is necessary before the body's defence mechanisms are overcome. Dengue, on the other hand, probably requires only a single infective mosquito-bite for its transmission. For this latter disease there is also the problem that no prophylactic drugs or vaccines are generally available for protective purposes.

MALARIA PROPHYLAXIS. The use of prophylactic drugs to combat malaria is well known. What is less widely appreciated is the fact that the use of these drugs has been increasingly complicated in recent years by changes in drug-resistance patterns amongst the malarial parasites. Since the late 1950s the most dangerous species *Plasmodium falciparum*, has become resistant, in one part of the world or another, to each group of drugs in turn. Resistance against proguanil* and pyrimethamine appeared first, followed in the early 1960s by resistance to chloroquine and amodiaquine (Peters 1982). In areas where chloroquine resistance is firmly established, the practice has been adopted of combining the traditional drugs with other substances. The principal combined drugs are Fansidar (pyrimethamine/sulphadoxine) and Maloprim (pyrimethamine/dapsone). However, these substances also have some limitations on their use, for example Fansidar is unsuitable for people who are sensitive to sulphonamide drugs (Ross Institute 1981).

It is clear therefore that the selection of suitable prophylactics for given destinations can no longer be made on a haphazard basis, and requires detailed medical advice. Evidence that the measures taken are frequently inadequate or inappropriate can be judged from the large number of tourists from Europe who currently contract malaria whilst visiting the tropics (Bruce-Chwatt 1982).

AVOIDING MOSQUITO BITES. The scope for tourists to reduce all these mosquito-related risks by minimising the number of bites sustained, has been generally underemphasised. The appropriate precautions of proven value are extremely simple. They involve choosing clothing which covers the arms and legs, and making routine use of insect repellents. These preparations are widely available as creams, gels or aerosol sprays and can be assumed when applied to the skin to have a working life in the tropics of about two hours before re-application is necessary.

Tropical airports

A problem too rarely mentioned in relation to mosquito hazards is connected with airports in the tropics. Most tourists assume that if they are taking their holiday in a malaria-free country, then the

---

* These drugs are also known under a variety of manufacturers' names: proguanil = Chlorguanide, Paludrine; pyrimethamine = Daraprim; chloroquine = Avlochlor, Nivaquine; amodiaquine = Camoquin, Flavoquine.

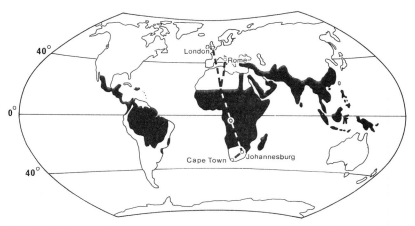

Fig. 6.4. Risk areas for malaria. The map also illustrates how travellers starting and ending their journeys in malaria-free areas may nevertheless be at risk at intermediate refuelling stops.

question of antimalarial protection need not arise. Unfortunately this can be an unwarranted assumption if the outward or return journeys include stopovers or even brief refuelling stops at incompletely protected airports in malarial areas (Breen 1963). The absence at many tropical airports of enclosed walkways or shuttle-buses to connect aircraft parking bays and transit lounges, may oblige passengers to walk considerable distances over the tarmac. At night this creates the opportunity for attacks by malarial mosquitoes especially if inadequate attention has been given to vector control within the airport perimeter (World Health Organisation 1972). Figure 6.4 illustrates journeys which begin and end in malaria-free areas, but where it would be prudent to take prophylactic precautions because of the risk of infection at the central African stopping point.

Tourism-related disease problems in temperate areas; the increased incidence of tick-borne diseases

The aggravation of disease hazards by the entry of tourists into unfamiliar habitats is a phenomenon which is not restricted to the tropics. In recent years the more extensive recreational use of previously remote temperate forest areas has increased the importance of a number of tick-transmitted diseases which were formerly restricted to occupational groups such as prospectors, trappers, foresters and farmers.

Tick-borne encephalitis in Europe and the USSR

It has been known since the 1930s that, in parts of Europe and the USSR, ticks can be involved in the transmission of virus diseases causing serious brain inflammation (encephalitis) in man. One such disease (Russian spring-summer encephalitis) first came to light during the exploration of the coniferous forests of Siberia and was shown to involve a high incidence of paralysis and a high mortality rate. Shortly afterwards a similar but milder form of encephalitis (Central European encephalitis) was identified from the forest areas of the Western USSR and Central Europe (Blaškovič 1967, Hoogstraal 1967).

Although traditionally these diseases have affected mainly foresters and agricultural workers, in recent years the problem has extended to tourists. Cases of encephalitis are now quite common amongst tourists visiting Austria, Czechoslovakia and Yugoslavia. In Brno, Czechoslovakia, 360 cases were recorded amongst holidaymakers over a period of four years, and in Yugoslavia almost 1500 suspected cases are reported annually (British Medical Journal 1978a). The most likely explanation for this increasing incidence is that many of the features now being incorporated into recreational facilities, such as forest camp sites and trails, represent very favourable habitats for ticks. In Europe, the important disease vector *Ixodes ricinus* (Fig. 6.5) has evolved the habit of congregating in forest-edge vegetation where there is a mixture of weeds, shrubs and isolated trees (Michalko 1967, Rehse-Küpper *et al.* 1978). This apparently gives the ticks the opportunity to position themselves on vegetation at different heights and thereby to make contact with a range of different-sized host animals (Nosek & Grulich, 1967). Apart from man, animals which serve as hosts for the ticks range from small species such as mice and hedgehogs to larger mammals such as deer (Fig. 6.5). This preference of the ticks for forest-edge vegetation is a crucial factor in bringing them into contact with tourists. Visitors from urban areas have the

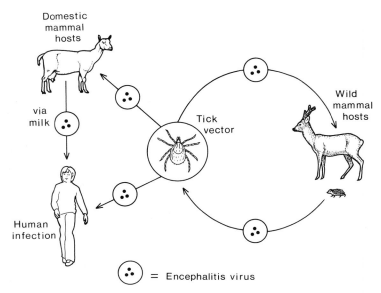

Fig. 6.5. Transmission cycle of Central European encephalitis. Wild and domestic mamals serve as hosts for the disease. Tourists can become infected when they are bitten by ticks or when they drink unpasteurised milk.

additional disadvantage that they are unlikely to have built up any immunity to the virus as a result of previous mild infections. Although *Ixodes ricinus* occurs in Britain and is responsible for transmitting louping-ill (a virus disease of sheep), in this country it does not transmit the encephalitis virus to man.

Little information is available about the relationship between encephalitis and recreational activities in the parts of the USSR affected by Russian spring-summer encephalitis. However, the family groups which frequently travel from towns to collect berries, edible fungi and pine cones in the forest are likely to be at risk. The activities organised in Young Pioneer Camps are also frequently forest-based and Pavlovsky (1966) describes an episode at a camp in the Kuznetsk Basin in which 30 cases of encephalitis and one death occurred amongst visiting children and staff.

Tick-borne diseases in North American recreational areas

In North America there has been a similar realisation that tick-borne diseases, previously regarded as occupational hazards amongst foresters, stockmen, and trappers, now warrant careful scrutiny in the context of recreation. The three diseases in question are Colorado tick fever, Rocky Mountain spotted fever and tick-borne relapsing fever.

COLORADO TICK FEVER. In the western mountains of North America, Colorado tick fever is transmitted to man as a result of bites from the wood tick *Dermacentor andersoni* (Fig. 6.6a). This species has similar habits to *Ixodes ricinus* in that it frequents shrubby vegetation for the purpose of attaching itself to its animal hosts. Mature ticks feed mainly on larger animals such as deer, mountain goats and bears, whereas immature ticks make use of smaller species such as ground squirrels and chipmunks. Humans pick up ticks when they come into contact with tick-infested vegetation alongside trails and around camp sites. If the tick is free of pathogens, then a bite is likely to cause only local irritation. If, however, it is carrying the virus, then an infection is likely to follow. Some hundreds of cases are now recorded annually from the western states of the USA and from British Columbia. Although in adults the infection usually involves only a mild fever, it is more serious in children and can cause encephalitis.

ROCKY MOUNTAIN SPOTTED FEVER. This tick-borne disease is so called because it was originally associated with the western mountains in North America and affected mainly pioneer explorers and settlers. It is usually characterised by a rash which starts on the hands and feet and subsequently spreads to the rest of the body. Its main area of occurrence has now moved from the western to the eastern half of the country, and increasingly, infections are associated with recreational activities. In 1977, for example, most of the 1115 cases (of which 42 were fatal) were concentrated in the eastern states of North Carolina, Virginia, Tennessee, Maryland and Oklahoma (British Medical Journal 1978*b*).

A common habitat for transmission to occur is in places where urban expansion has caused farmland to be abandoned, resulting in the development of the shrubby vegetation favoured by ticks (Burgdorfer 1975). Dogs being exercised by local residents in such

107

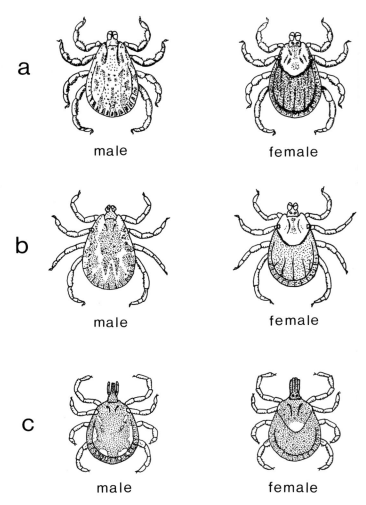

Fig. 6.6. Hard-bodied ticks involved in disease transmission in recreational situations in North America: (a) wood tick (*Dermacentor andersoni*); (b) dog tick (*Dermacentor variabilis*); (c) lone star tick (*Amblyomma americanun*) (based partly on Arthur 1962).

situations are liable to pick up dog ticks (*Dermacentor variabilis*) (Fig. 6.6b) a proportion of which carry the disease organism. Human infections can then occur if the dog's owner removes and crushes an infected tick during the process of 'deticking' the animal. This allows the pathogens to enter the body through abraded skin or if the eyes are rubbed with contaminated hands. Additionally, live ticks may transfer themselves from the dog to its owner. The central role of dogs in this transmission chain is supported by the findings of a survey in Mississippi which showed that 35 out of 38 sufferers from spotted fever either owned or had close contact with dogs, and seven of them had deticked dogs prior to illness.

In other situations, such as the camp and picnic sites established in the oak-hickory forests of the Ozark plateau, another tick, *Amblyomma americanum*, is the important disease transmitter. This is the 'lone star tick' so named because the conspicuous pale mark on the back of the female resembles a cattle brand (Fig. 6.6c).

TICK-BORNE RELAPSING FEVER. The third tick hazard encountered in North America is associated not with outdoor situations, but with the interior of log cabins and other simple wooden buildings now widely used for recreational purposes. In these situations the soft tick, *Ornithodoros hermsi*, can transmit relapsing fever to man. Animals such as squirrels and chipmunks, which have established themselves in cabin roofs and floors serve as reservoirs for the disease (Fig. 6.7). In one well documented episode from Spokane County, in Washington State, ten out of twenty members of a scout troop who slept overnight in some dilapidated cabins contracted the infection (Thompson *et al.* 1969). Examination of the cabins showed that rodents were nesting in the roof and under the floor space and that infected *Ornithodoros hermsi* were present in the walls. A more extensive outbreak involving 62 people in the Grand Canyon National Park, Arizona, in 1973 was similarly found to be associated with rodent- and tick-infested rustic cabins (Pratt & Darsie 1975). Relapsing fever is a prolonged and debilitating disease involving recurrent bouts of fever extending over a period as long as ten weeks.

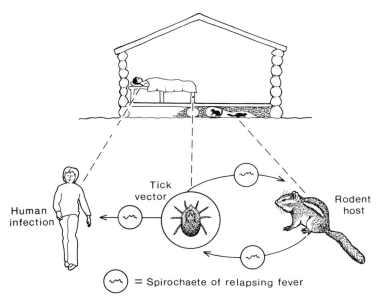

Fig. 6.7. Transmission cycle of tick-borne relapsing fever. Rodents in holiday cabins serve as hosts to the disease which is transmitted by night-time bites from the soft tick *Ornithodorus hermsi*.

Control measures directed against ticks in recreational areas

Some success has been achieved in reducing the incidence of the lone star tick in recreational areas by modifying the vegetation. This involved removing a proportion of the overhead shade trees to let in more sunlight, and cutting the lower vegetation to produce a short grass cover. Both these measures apparently make the habitat less attractive to ticks (Mount 1981). A complementary method has been to apply acaricides (chemical control agents used against acarines such as ticks and mites) in the form of chemical sprays and dusts (Mount 1983). Whilst these techniques are applicable on a small scale they are not practical for large areas. In Europe where recreational facilities are often close to agricultural land, there is also the risk of acaricides endangering domestic stock, bees and other pollinating insects.

Control of wild mammal hosts is not usually a practical or publicly acceptable approach to the problem, although in the case of relapsing fever, rodent-proofing of cabins will help break the transmission cycle. In Europe there is some merit in dissuading children from adopting hedgehogs as camp site pets because of the risk of these serving as encephalitis reservoirs. A vaccine is available for use against Rocky Mountain spotted fever for persons particularly at risk, but the protection given is of relatively short duration and booster doses are required every 12 months. The use of vaccines to give protection against encephalitis is more problematical and carries a high risk of adverse side effects.

In view of these complications, managers of recreational facilities should play an active part in advising visitors about measures they can take to protect themselves from tick bites. These include dressing to reduce tick access, applying repellents to clothing, and searching the body for ticks at least twice a day. This last recommendation is based on the fact that ticks usually wander about on the skin for some hours before settling down to feed and there is considerable merit in trying to intercept them before this happens.

The contribution of ecology is readily apparent in situations such as these, where the habitat preferences and behaviour patterns of host and vector species are important considerations. However, a comprehensive approach to environmental health also needs to take account of diseases which are contracted by more direct routes. A topical issue in this latter area concerns the health implications of bathing in sewage-polluted waters.

Health implications of bathing in sewage-polluted waters in temperate regions

It is timely to re-examine this issue not because of a change in recreational patterns, but rather because new information has become available about the health risks which might be involved.

In most Western countries the rapid growth of towns and industries during the industrial revolution led to the widespread discharge of sewage into inland and coastal waters. Ironically amongst the sites affected were the coastal bathing resorts which had been developed earlier to capitalise on the supposed therapeutic effects of sea bathing, and which then continued to expand with the demand for coastal recreation. Streams and rivers were also widely contaminated by the use of 'storm-water overflow' waste-disposal systems. These systems had the dual function of carrying away both storm-water and sewage and were designed to discharge into local watercourses only when overloaded during periods of heavy rain. Many are still in use today but the increased volume of sewage being carried causes them to overflow into rivers more frequently.

The knowledge that diseases such as typhoid fever and bacillary dysentery can be contracted from faecally contaminated drinking water carries with it the implication that people bathing in polluted water might be similarly at risk. Even expert swimmers are known to swallow at least 10 ml of water every time they bathe, and inexpert ones would be expected to take in very much more. Despite the logic of this argument, it has proved extremely difficult to arrive at a realistic assessment of the risks involved. One of the difficulties is that infections contracted whilst bathing are liable to be masked by similar infections transmitted by other routes such as person-to-person contact, or via contaminated food and drinking water.

In Britain, official attitudes to the problem were greatly influenced by the deliberations of a working group set up by the Public Health Laboratory Service (Moore 1959). This group reached the conclusion that 'a serious risk of contracting disease through bathing in sewage-polluted seawater is probably not incurred unless the water is so fouled as to be aesthetically revolting'. This view was greeted with some relief by the officers of local authorities anxious to avoid further expenditure on waste-disposal systems. It conflicted, however, with earlier observations made by American workers which pointed to a clear relationship between the level of sewage pollution and the incidence of gastrointestinal disorders amongst bathers, even in conditions where obvious signs of pollution were absent (Stevenson 1953). Information which has accumulated since this period suggests that the American view came nearer to the truth.

Since the 1950s sewage pollution has been firmly implicated in a number of outbreaks of disease. For example ten cases of typhoid fever amongst bathers at a coastal beach in Western Australia proved to be associated with a broken sewer pipe (Public Works 1961), and 31 cases of bacillary dysentery amongst bathers at Dubuque in Iowa were shown to be associated with sewage pollution of a river bathing site on the Mississippi River (Rosenburg *et al.* 1976).

Using a different approach, a survey carried out in 1971 by the consumer associations of France and Belgium concluded, on the basis of 9000 replies, that the chances of becoming ill after a coastal holiday were approximately doubled if the holidaymaker bathed in the sea. Whilst some of the disorders, such as ear infections and sinusitis, which showed an increased incidence in bathers, could have been produced by the bathing *per se*, irrespective of the cleanliness of the water, the other infections identified were very likely to have been associated with sewage pollution. For example, there were 18 cases of paratyphoid, 23 cases of hepatitis and 252 cases of gastroenteritis amongst bathers, whereas the comparable figures for non-bathers were 3, 0 and 32 respectively.

Most recently, an extensive study of marine and brackish water beaches carried out by the United States Environmental Protection Agency (Cabelli 1979, 1981) has established a general relationship between the level of sewage pollution and the incidence amongst bathers of illnesses involving gastrointestinal symptoms (Fig. 6.8).

These findings give strong support to the view that infections of the kind listed in Table 6.1 can be contracted by bathing in sewage-polluted water.

The use of indicator organisms

Technical complications arise when one proceeds to the obvious next stage of attempting to define and enforce water quality standards. Sophisticated laboratory facilities and special growth media would be necessary to test comprehensively for all the pathogens listed in Table 6.1. These facilities would not usually be available for routine monitoring programmes. Consequently reliance is placed on standardised tests for organisms which are normal inhabitants of the human alimentary tract and which can be expected to occur in abundance in sewage-polluted waters. The assumption being made is that their presence indicates the possibility of pathogens being present. It will be noticed in the results shown in Figure 6.8 that the bacteria known as faecal streptococci or enterococci are being used in this role as 'indicators' of the level of sewage pollution.

Table 6.2 gives examples of bathing water standards which have been proposed using various indicator groups. It will be apparent from this table that even using the same organisms considerable differences

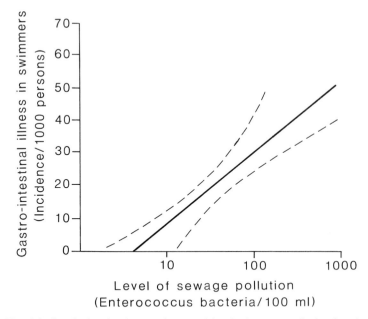

Fig. 6.8. Graph showing how an increased level of sewage pollution in a bathing water increases the likelihood of gastrointestinal illnesses amongst swimmers (broken line indicates 95% confidence levels) (after Cabelli 1981).

of opinion exist about what constitutes an appropriate standard. These variations result largely from the still meagre amount of information available about the precise relationships between indicator organisms and true pathogens. It is to be expected that these would differ from one environmental situation to another. For example, in some localities pollution from animal wastes could increase the total coliform count without necessarily indicating an increased human health hazard. Conversely in a community with a typhoid carrier, typhoid organisms could be present in bathing water which registered a relatively low coliform count. Whilst such complications may call into question some of the bathing water standards currently being used, the case for standards remains a strong one. Probably the most appropriate future development would be for indicator systems to be devised which incorporate calibration factors for local conditions.

Table 6.1. *Diseases potentially contractable
from sewage-polluted bathing waters in temperate regions*

Disease	Causative organism
Gastroenteritis	(i) *Escherichia coli* (enteropathogenic)
	(ii) *Campylobacter fetus*
	(iii) Norwalk virus
	(iv) Rota virus
Bacillary dysentery	*Shigella* spp.
Typhoid fever	*Salmonella typhi*
Paratyphoid fever	*Salmonella paratyphi*
Viral hepatitis	Hepatitis A virus
Poliomyelitis	Polio virus

Table 6.2. *Examples of bathing water standards
based on indicator organisms*

Country/Agency, etc.	Indicator	Standard (upper limit/ 100 ml)
European Economic Community	Total coliforms	10 000[a]
USA – 23 States, USSR	Total coliforms	1 000
European Economic Community	Faecal coliforms	2 000[a]
USA – 12 States	Faecal coliforms	200
European Economic Community	Faecal *Streptococci* (= *Enterococci*)	100[b]

a 95% of the samples must be equal to, or lower than, value specified.

b 90% of samples recommended to be lower than this value.

Warm-water amoebae

Although in general the heated and chlorinated waters of municipal bathing pools can be regarded as entirely safe alternatives to natural waters, there is at least one potentially dangerous organism which can thrive even in these conditions. This is the protozoan *Naegleria fowleri*

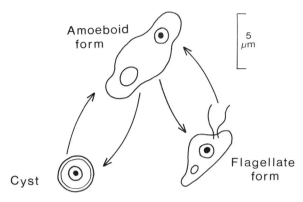

Fig. 6.9. Diagram illustrating the various life forms of the protozoan *Naegleria fowleri*. The amoeboid form can be responsible for cases of meningo-encephalitis amongst swimmers using heated bathing waters.

(Fig. 6.9). It usually lives harmlessly in soil or water, but occasionally the amoeboid form enters the nostrils of a bather and migrates along the olfactory nerve to the brain. Here it is liable to cause meningo-encephalitis, an inflammation of the brain and surrounding membranes which is usually fatal (Jadin 1981). In a notable episode from Czechoslovakia sixteen swimmers contracted meningo-encephalitis from a heated, chlorinated pool. It transpired subsequently that amoebae had been able to maintain themselves behind a damaged section of the pool wall (Cěrva & Novǎk 1968, Kadlec *et al.* 1978). Virtually all such outbreaks have been associated with warm waters in temperate regions (Willaert 1974) and this agrees with the laboratory observation that *Naegleria* colonies grow freely at temperatures as high as 45 °C (Griffin 1972). Whilst statistically the overall hazard from *Naegleria* is minimal, the occurrence of meningo-encephalitis cases amongst the users of a swimming pool constitutes a strong argument for taking it out of commission.

Communication deficiencies in relation to recreational health hazards

The disturbing conclusion to be reached from this short review of recreational health hazards is that participants are frequently unaware of even the elementary precautions they should take for their own protection.

In most Western countries literature is available which provides advice to holidaymakers before they set out on their travels. In Britain the relevant official publication takes the form of a 16-page booklet entitled '*Protect Your Health Abroad*' produced by the Department of Health and Social Security and available from Health Departments (Department of Health and Social Security 1985). This provides information on the range of vaccinations formally required, or otherwise recommended as desirable, when visiting particular countries. It also stresses the need to seek medical advice about malaria prophylaxis. Detailed advice about appropriate health safeguards for travellers is also provided in publications from the Ross Institute (1980) and the British Medical Association (Walker & Williams 1983). Unfortunately little of this information is incorporated into travel brochures, nor do travel agents habitually draw the DHSS booklet to their clients' attention or show any working knowledge of its contents (Holiday Which, 1983).

Having arrived at his or her destination, the tourist could often benefit from simple on-site advice, for example about the need to remove ticks or avoid waters occupied by schistosome-carrying snails. This could be provided most effectively by tour guides or the managers of recreational facilities. Whilst some organisations, such as the National Park Service in the United States, have an excellent record in this respect, many commercial enterprises provide no useful guidance on these matters.

Finally, the tourist who becomes sick after returning home may suffer unnecessarily because his local doctor fails to recognise an unfamiliar illness. For example there have been numerous instances where the treatment of malaria has been delayed, sometimes with fatal consequences, because it was initially mistaken for influenza (Maegraith 1963, Breen 1963).

Tourism is properly recognised as providing people with the opportunity to encounter new situations and enjoy new experiences. These benefits are obviously negated if the traveller becomes ill in the process.

7.
INSECT
NUISANCES

Insects can interfere with recreational activities in a variety of ways quite apart from the threat of transmitting disease. Visitors may be stung by wasps, bitten by various flies, ranging from horseflies to sandflies, or have their enjoyment spoilt by dense enveloping swarms of non-biting species such as chironomid midges and mayflies. Although in some circumstances these incidents are of a trivial nature, there are situations where the amount of annoyance caused is so great that the continued use of a recreational area is jeopardised.

Mass swarms of non-biting insects

The swarming insects most likely to cause problems in recreation areas are the non-biting midges (Chironomidae and Chaoboridae), the mayflies (Ephemeroptera) and the caddis flies (Trichoptera) (Fig. 7.1). Although none of these groups includes biting species, the swarms they form are often so dense that it is difficult to avoid inhaling the insects or to prevent them flying into one's eyes. Moreover, sensitised individuals can develop serious allergic reactions to the dust-like debris blown from the heaps of accumulated insect bodies and cast skins (Frazier 1969).

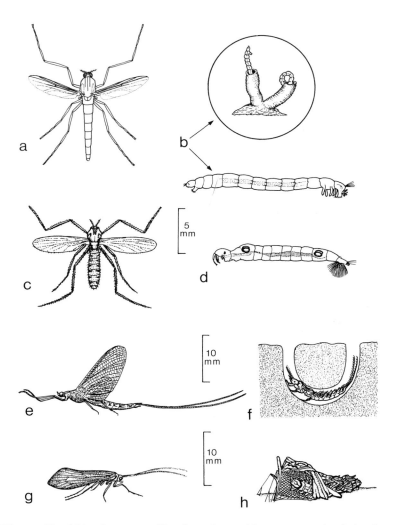

Fig. 7.1. Non-biting insects capable of creating problems at recreational sites by appearing in dense swarms. (a–b) Chironomid midge, adult and larva, inset shows structure of larval tubes. (c–d) Chaoborid midge, adult and floating transparent 'phantom larva'. Unlike a mosquito (Fig. 6.3) neither of these types of midge has a proboscis. (e–f) Mayfly (*Hexagenia*), adult, and larva shown occupying its U-shaped tube. (g–h) Caddis fly (Hydropsyche), adult, and larva shown occupying its shelter and catching net. (Diagrams based partly on Harwood & James 1979, Wallace & Merritt 1980).

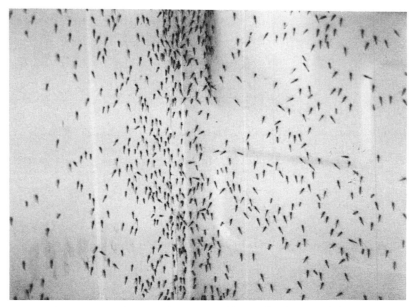

Fig. 7.2. Chironomid midges resting on the side panels of a lakeside house in California (photograph by courtesy of Dr M. S. Mulla, University of California, Riverside).

Chironomid midges

Of the various groups in this category, the chironomid midges (Figs. 7.1a, 7.2) are of most widespread importance and have caused problems in North and Central America, Africa, Europe, Japan and New Zealand (Grodhaus 1975). A typical example of the difficulties resulting from chironomid swarms is described by Lewis (1956) from the Sudan

Chironomidae were a serious pest at the Nile Hotel causing intense annoyance to guests and staff. In the mornings dead midges were swept up by the bucketful. Nearly all light had to be extinguished in the evenings. Many people could not sleep owing to allergic effects, and a considerable proportion of the staff spent periods in the asthma camp. The pest was so bad that removal of the town was discussed.

The early development of chironomid midges takes place in a variety of stagnant and flowing-water habitats. The larvae live in silk or mud tubes which they construct on the leaves of aquatic plants or on the bed of a river or lake (Fig. 7.1b). They feed on organic particles and microscopic algae and are therefore particularly abundant in waters which are naturally productive, or ones which have been enriched by human or agricultural wastes.

Unfortunately, designers of recreational facilities often create problems for their clients by unwittingly producing just the environmental conditions which chironomids prefer. In southern California for example, the artificial lakes which have been incorporated into some high-quality residential and recreational developments are warm, shallow and rich in nutrients and represent ideal breeding sites for chironomids (Fig. 7.2) (Grodhaus 1963, Mulla 1974). At these sites, massive swarms of midges are liable to interfere with outdoor activities such as boating and fishing, and where the insects are attracted indoors by house lights, they often cause unsightly stains on walls and furnishings.

Chaoborid midges

Similar problems but on a more localised scale can be caused by members of the other group of midges, the chaoborids. Whilst the adults closely resemble chironomids (Fig. 7.1c), the larvae are quite different and take the form of remarkable transparent floating organisms, usually referred to as 'phantom larvae' (Fig. 7.1d). On some African lakes dense swarms of adult chaoborids are credited with causing the suffocation of fishermen. In temperate regions the best example of a troublesome species is provided by the 'Clear Lake gnat' (*Chaoborus astictopus*). At Clear Lake in northern California swarms of this insect have been recognised as a serious nuisance to fishermen since the 1930s (Cook 1965).

Control of midges

Early attempts to control the Clear Lake gnat using the organochlorine insecticide DDD (dichloro-diphenyl-dichloroethane) provided a classical demonstration of the hazards of introducing pesticides into lake food chains. Pesticide residues were shown to have accumulated in the lake's fish population and also to have been passed on to fish-eating birds such as grebes, many of which died as a result (Hunt & Bischoff

1960). The subsequent use of an organophosphorus insecticide appeared to involve fewer side-effects but eventually proved ineffective because of the development of insecticide-resistance in the midge population. Most recently, reliance has been placed on a biological control method, using a fish, the Mississippi silversides (*Menidia audens*), introduced into the lake specifically for this purpose (Cook 1981).

Organophosphorus insecticides have been used with some success against chironomid midge larvae in the southern Californian lakes. Here too, however, there are indications of developing resistance and adverse side effects on fish food organisms (Ali & Mulla 1978*a*, *b*, 1979). At these sites use has also been made of growth-regulating chemicals. By retarding larval and pupal development these substances may provide a few weeks respite from the midge nuisance. Like the insecticides, however, they can also have adverse effects on other organisms (Ali & Mulla 1978*b*). In view of the problems involved in trying to control midges in shallow recreational lakes, one is bound to question the wisdom of constructing some of these facilities in the first place.

Mayfly and caddis fly swarms

Many of the problems caused by swarming mayflies and caddis flies also seem to have their origins in modifications in the habitat made by man, although in these cases not necessarily ones made for recreational purposes. In the upper reaches of the Mississippi River swarms of the mayfly *Hexagenia* (Figs. 7.1e, 7.3) are particularly associated with localities where the river has been impounded for power generation. The accumulation of fine silt and organic matter behind these impoundments provides favourable conditions for the mayfly larvae, which occupy U-shaped burrows in the river bed and feed by ingesting organic particles (Fig. 7.1f). It has been estimated that a single impoundment can support more than a thousand million larvae (Carlander *et al.* 1967). When the mayflies reach maturity and emerge as adults the resultant swarms can interfere seriously with the public use of riverside picnic areas, boat-docking facilities and lighted baseball parks (Fremling 1968). Roads on bridges can become buried by mayfly bodies to such a depth that snowploughs are required to

Fig. 7.3. A swarm of *Hexagenia* mayflies on a Mississippi River bridge at Winona, Minnesota, July 1966 (photograph by courtesy of Dr C. R. Fremling, Winona State University, Minnesota).

clear the obstruction (Fig. 7.3). The masses of bodies are not only evil-smelling but are also liable to cause allergic reactions in sensitised individuals (Figley 1940).

The caddis flies which produce troublesome swarms are mainly species whose larvae live in flowing water and collect their food by trapping plant and animal fragments in silk nets (Fig. 7.1g,h). Such food supplies are abundant in the outflows of lakes, particularly ones which have experienced nutrient enrichment or 'eutrophication' from the input of sewage, industrial wastes or agricultural fertilisers. One such situation is where the Niagara River flows out of the much enriched Lake Erie on the United States/Canada border. Here dense swarms of the caddis fly *Hydropsyche* can create a serious nuisance in urban areas along the riverside (Munroe 1951, Peterson 1952).

Although in theory it would be possible virtually to eliminate these organisms from critical sections of large rivers by using repeated doses

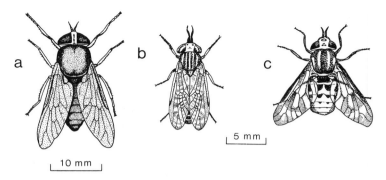

Fig. 7.4. Examples of the larger biting flies shown in their normal resting positions. (a) Black horsefly (*Tabanus atratus*) a widespread species in North America; (b) common cleg (*Haematopota pluvialis*, recognisable from its mottled wings, a major nuisance in open woodland areas in southern Britain; (c) deer fly (*Chrysops discalis*), deer flies have characteristically spotted abdomens and banded wings, and may interfere with recreational activities in both Europe and North America. (Diagrams based partly on Harwood & James 1979).

of insecticide, the costs involved and the likelihood of adverse side-effects rule this out as an acceptable strategy. In such circumstances the most appropriate response from recreational planners is to discourage the development of major riverside facilities in vulnerable areas.

Problems in recreational areas caused by biting flies

A large range of biting flies can cause problems in recreational areas. The list includes mosquitoes, horseflies, clegs, deer flies, stable flies, biting midges, blackflies and sandflies. With some species it has not been possible so far to define larval habitats with sufficient accuracy to mount effective control measures. This is the case, for example, with the tiny biting midges or 'punkies' (*Culicoides*) which can be extremely troublesome in North temperate regions. These insects seem to be able to pass their early stages in almost any form of wet peaty soil. Similarly the large biting species such as horseflies (*Tabanus*), clegs (*Haematopota*) and deer flies (*Chrysops*) (Fig. 7.4a–c)

124

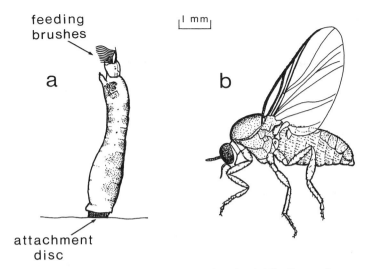

feeding
brushes

a

b

attachment
disc

Fig. 7.5. Larva (a) and adult (b) of blackfly (*Simulium*). The fly may be recognised from its humped-backed appearance and short, broad wings.

appear to be able to utilise a very wide range of damp habitats for larval development.

Much more progress has been made in understanding the ecology of blackflies and sandflies and this information deserves careful examination in connection with recreational developments.

Blackflies (*Simuliidae*)

Like net-spinning caddis larvae, the immature stages of blackflies feed on organic particles, although of a much smaller size than those used by caddis. The filtration mechanism consists of a pair of brushlike structures on the head (Fig. 7.5a), and a rapid flow of water is necessary for this apparatus to work efficiently. This method of feeding results in blackfly larvae being particularly abundant in the outflows and spillways from productive lakes and in the swift-flowing reaches of nutrient-rich rivers and streams. Consequently, it is in the vicinity of these habitats that the biting adult blackflies (Fig. 7.5b) are most likely to make a nuisance of themselves. For example in

North America, blackfly problems in Los Angeles County occur in areas close to lake outflow and dam spillways which support large populations of larvae (Pelsue *et al.* 1970, Hall 1972). Similarly in New Hampshire, blackflies in a recreational area were found to be derived from larvae living in the outlet conduits of artificial ponds and lakes (La Scala & Burger 1981). In Southern England the high incidence of blackflies along the River Stour seems to be associated with the nutrient enrichment of this river from agricultural sources. This causes an abundant supply of particulate organic matter to become available to the larvae (Hansford & Ladle 1979).

Blackfly bites are painless initially but in a few hours usually produce inflamed, itching weals. In sensitised individuals or those receiving multiple bites, more generalised allergic reactions can occur, including swollen limbs, headache and fever. In the vicinity of the River Stour 600 people were reported to have required medical attention for blackfly bites in one four-week period. Fortunately, the tropical areas where blackflies can transmit the causative organisms of river blindness (onchocerciasis) have not yet been significantly developed for tourism. This involvement of blackflies in transmission of disease has, however, provided an important additional stimulus for investigating possible control measures.

Control of blackflies

The mass control of blackfly larvae raises similar problems to those encountered in relation to midge control. The insecticide DDT has been known since 1943 to be effective in killing blackfly larvae and was used widely for this purpose during the 1950s in recreational areas in North America. Subsequent detailed studies showed that harmless invertebrate animals were also affected and that the pesticide was likely to interfere with the breeding of lake fish (Burdick *et al.* 1964). These findings echoed a similar concern being expressed about the agricultural use of DDT and have led to a ban on its use in blackfly control programmes in most parts of the world. Attention then turned to organophosphorus insecticides and these have been shown to be effective against blackflies, without having such extreme side-effects as DDT (Wallace & Hynes 1981). In the long run, however, their usefulness is likely to be diminished by the blackflies becoming resistant.

A comprehensive search for biological control agents (Laird 1981) has shown that a bacterium, *Bacillus thuringiensis*, has considerable possibilities in this role. This organism can be cultured on a large scale and has been used for many years in control programmes directed against forest caterpillars. Recent experiments with a particular strain (*israelensis*) have shown that, when introduced in suspension into a blackfly-infested stream, it will cause appreciable mortality (Molloy & Jamnback 1981). The bacteria act by damaging the gut wall of the larvae (Lacey & Federici 1979). This control method seems to have the additional advantage of producing minimal effects on non-target organisms. It now remains to be seen whether a way can be found, possibly by means of chemical encapsulation, of improving the efficiency of dispersal of the bacterium in stream systems.

Finally, in some circumstances, physical alteration of the habitat can play a part in blackfly control. Where impoundments are constructed on small watercourses it is often possible, by means of siphoning arrangements, to dispense with the permanently water-covered spillways which would otherwise provide ideal settlement areas for the larvae (Addison 1964).

Ceratopogonid sandflies

In warmer parts of the world, beaches at coastal resorts can be rendered virtually unusable by the activities of biting sandflies. These insects are closely related to the midges and 'punkies' of damp habitats inland and are also included in the family Ceratopogonidae. Confusingly the term sandfly is also applied to insects in the subfamily Phlebotominae, which are responsible for the transmission of leishmaniasis in the tropics and subtropics. Ceratopogonid sandflies pass their early stages either in beach sand or in the mud of tidal mangrove swamps (Fig. 7.6). The sand-breeding species, such as *Leptoconops*, tend to bite during the daylight hours, with a peak of activity in the afternoon, whereas the mud-breeding species such as *Culicoides* are usually active at night, with peaks of activity at dawn and dusk. Holidaymakers who unwittingly choose a hotel close to both types of breeding site may render themselves vulnerable to sandfly attack at whatever time of the day or night they venture out of doors. Although sandfly bites are scarcely detectable at the time, they usually cause intense irritation after a few hours and the resulting discomfort can

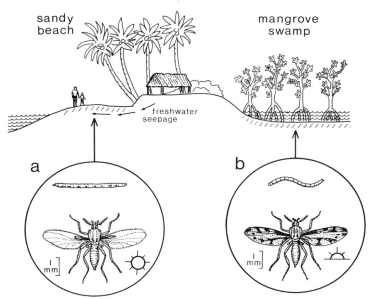

Fig. 7.6. Breeding sites of sandflies in coastal regions of the tropics and subtropics. (a) *Leptoconops* larvae develop in moist beach sand, and the adult flies have milky-white wings; (b) *Culicoides* larvae develop in the mud of mangrove swamps, and the adult flies have spotted wings.

last for days. In general it is considered that an average person will tolerate a maximum of about five sandfly bites an hour or 30 bites a day (Linley & Davies 1971). When this figure is compared with the 3000 bites an hour which could theoretically be sustained by an unprotected human subject on a beach lacking control measures, the magnitude of the problem becomes apparent. Areas where major sandfly problems have been reported include the Caribbean, Florida, the Gulf Coast of the United States (Linley & Davies 1971), the Seychelles (Laurence & Mathias 1972) and Australia (Kettle *et al.* 1979).

Failure to appreciate the nuisance potential of sandflies can lead to recreational developments being designed in a form which actually aggravates the problem. Along the southern Queensland coast of Australia, the construction of a complex system of canals and mooring quays has greatly increased the area of substratum suitable for sandfly development (Kettle *et al.* 1979). Likewise in the Caribbean, the removal of sand from beaches for constructional purposes has exposed sand layers which are moistened by freshwater seepages. These areas provide excellent breeding sites for *Leptoconops* sandflies and in some instances a nuisance problem has been created where none existed before (Linley & Davies 1971).

Control of sandflies

A number of the countermeasures theoretically available for use against sandflies prove to have major drawbacks in practice. In tropical conditions, repellent creams and aerosols applied to the skin can lose their efficiency in less than two hours because of increased rates of absorption, volatilisation and loss due to sweating. This means that as a defence against sandflies on a tropical beach, repellents need to be reapplied to all exposed skin areas at two-hourly intervals and also after each bathing session. Moreover preventing the flies from biting does not stop them from causing irritation by crawling on the skin, and flying into the eyes and nostrils.

Although widespread spraying with insecticides has sometimes been used as an emergency control measure the expense of treating large areas of sand or saltmarsh is an economic argument against this as a regular procedure quite apart from the risk of causing general environmental contamination.

HABITAT MODIFICATIONS. The remaining possibility, and the one which shows most promise, especially in dealing with the mud–dwelling *Culicoides* species, is to modify the habitat to make it unsuitable for sandfly development. *Culicoides* sandflies are especially adapted to utilise mud areas which are flooded periodically and shallowly by the tide (Fig. 7.7a). By a special arrangement of dikes, pipes and tide gates (Fig.7.7b) it is possible to flood the mud permanently with seawater and thus make it unusable. This method has been successful in Jamaica, the Panama Canal Zone and in Florida, where reductions in fly populations as great as 95% are claimed (Linley & Davies 1971).

Benefits of a less dramatic nature can also be obtained by modifying sand habitats to make them unsuitable for *Leptoconops* sandflies. These species favour hollows which are moist by virtue of being close (within about 40 cm) to the underlying water table. By filling such

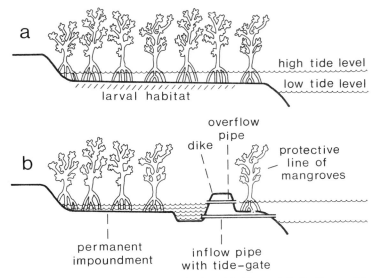

Fig. 7.7. Measures designed to destroy the larval habitats of *Culicoides* sandflies. The construction of a system of dikes and pipes causes the mangrove swamp to be permanently flooded with seawater. In these conditions the fly larvae cannot survive (based on Linley & Davies 1971).

hollows with extra sand it is often possible to keep the surface layers sufficiently dry as to make them uninhabitable (Linley & Davies 1971).

Stable flies

The other important biting-fly problem associated with coastal beaches in warmer areas is that caused by the stable fly (*Stomoxys calcitrans*) (Fig. 7.8). This species is so named because it has been traditionally associated with stables, where its larvae develop in manure and organic debris. Its presence on beaches is also linked with organic matter, but in this case with the heaps of rotting seaweed and sea grass (*Zostera*) which tend to accumulate along the tide line. The presence of a few hundred of these blood-feeding flies on a beach can cause it to be abandoned by holidaymakers (Hansens 1951). Infestations along the Gulf Coast of Florida, where the problem can be particularly severe, have been estimated to cost the local tourist industry a million dollars a day in lost revenue (Newson 1977).

5 mm

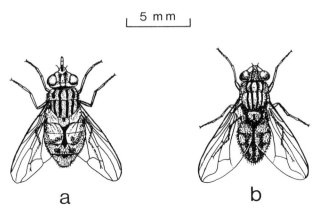

a   b

Fig. 7.8. (a) Stable fly (*Stomoxys calcitrans*), a biting species which can cause serious problems on beaches in southern parts of the USA. Stable flies resemble houseflies (*Musca domestica*) (b), but differ from them in having a more robust body and projecting mouthparts (based partly on Zumpt 1973).

Although insecticide sprays have traditionally been used against fly larvae occupying strandline debris, a sounder approach from an environmental point of view is to remove the debris physically from prime beach areas.

Problems associated with stinging insects

Wasps

Of the stinging insects, which include hornets, wasps, bees and some ants, usually only the wasps or 'yellowjackets' (Fig. 7.9a) warrant special attention in recreational areas. The reason for their importance in this context is the habit they have developed of exploiting the foodstuffs brought by visitors to picnic and campsites. With some species this foraging even extends to the meat laid out on picnic plates or being cooked on barbecues. Attempts to swat the insects often result in people being stung, and whilst in non-sensitised persons a sting produces only local pain, swelling and redness, disappearing in a few hours, in susceptible individuals fatal shock reactions can develop (Barr 1974). In the United States alone, wasps are recorded as causing

131

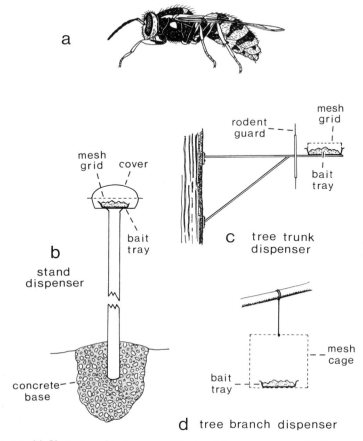

Fig. 7.9. (a) *Vespa pensylvanica*, one of the troublesome wasp species frequently encountered at picnic sites in North America. (b–d) various methods of presenting chemical baits used for wasp control; mesh grids allow access by wasps but not birds or rodents (based on Grant *et al.* 1968).

15–20 deaths each year and this is probably a considerable underestimate. Most wasp problems are associated with the *Vespula vulgaris* group of species (Table 7.1). Not only do these species visit picnic sites regularly but also they frequently construct their nests underground where they are difficult to find and destroy. Other species such as those in the *Vespula rufa* group may actually serve a beneficial function by consuming defoliating caterpillars.

132

Table 7.1. *Distribution of wasps in the* Vespula vulgaris *group* (*from Akre & Davis* 1978)

	North America	Europe	Asia (and Japan)
Vespula germanica[a]	× (eastern)	×	
Vespula lewisi			×
Vespula maculifrons	× (eastern)		
Vespula pensylvanica	× (western)		
Vespula vulgaris	× (transcontinental)	×	×

a Introduced into Chile, New Zealand, Tasmania, Australia and South Africa.

CONTROL MEASURES. In North America, which has an above-average quota of troublesome wasp species, much attention has been given to the possibility of controlling wasps with chemically impregnated baits. The principle of this technique is to induce the insects to take a bait containing an insecticide or growth hormone (Fig. 7.9b–d). When the bait is carried back to the nest to be shared with the larvae and the queen, it eventually causes the destruction of the colony. In one such trial in California, fish-based cat food mixed with an insecticide was used as the bait (Ennik 1973). Wire cages containing this material were hung up in trees in the recreation area. In addition tree trunks and branches were splashed with heptyl butyrate, one of a number of chemicals known to be attractive to wasps (Davis *et al.* 1969). The wasps carried the impregnated bait back to the nest and after two days baiting this produced a reduction of insects in the test area of between 75 and 95 %. A resurgence of the problem which occurred after 10 days was probably caused by the emergence of adults which had been present in the colony as pupae during the baiting and had thus avoided the insecticide. A second treatment was sufficient to deal with these remaining individuals.

These techniques have been most successful when used against the western species *Vespula pensylvanica*. In the eastern and southern States the wasp species appear to react less readily to attractants and the situation can arise where only a few foragers take the bait and are liable to be killed themselves by the insecticide before they have had an opportunity to carry a damaging dose to the colony. For these

133

situations, incorporation of growth-regulating substances rather than insecticides into baits has been tried (Parrish & Roberts 1983). These substances control colony expansion by interfering with the maturation of pupae.

With all chemically based control methods there is a risk of non-target species being affected. In Britain during the 1960s, a wasp bait containing a chlorinated-hydrocarbon insecticide in a sugar base was implicated in the poisoning of honeybee colonies (Spradbery 1973). Honeybees are more efficient than wasps in communicating information about rich food sources to the rest of the colony and therefore are more likely to suffer serious ill-effects if the food source contains a toxic agent. The use of growth-regulators for wasp control involves similar hazards if the substance is presented in sugar or honey baits which might be taken by honeybees, rather than in protein-based material (Zdarek *et al.* 1976). When devising control programmes attention must also be given to the possibility of causing more damage to useful members of the *Vespula rufa* group than to the troublesome *Vespula vulgaris* species (MacDonald *et al.* 1976).

Insect nuisances in perspective

It is difficult to avoid the conclusion that recreational planners frequently give insufficient attention to the potential impact of insect nuisances on their enterprises. Failure to anticipate such problems can result in major additional expenditure on attempted control measures or lead to a loss of revenue from underused facilities. Moreover hasty and ill-conceived remedial actions are frequently a cause of general environmental damage. It is particularly unfortunate that some recreational schemes unwittingly include features which aggravate insect problems rather than minimising them. It should also be recognised that there are some sites, both natural and man-made, with such intractable nuisance problems, that for all practical purposes they can be ruled out as places for organised recreational development.

8.
HAZARDS ASSOCIATED WITH LARGER ANIMALS

A serious difficulty in attempting to reduce the hazards posed by larger animals, is not that the public is unaware of the species concerned, but rather that their image of these creatures is often inaccurate. Folklore, travellers' tales and sensationalised media treatment have all contributed to the creation of false stereotypes which differ in important respects from biological reality. Whilst such images may serve to excite and entertain, they rarely provide a sound basis for practical management. The routine overstatement of an animal hazard may lead not only to unnecessarily repressive measures being directed against the species concerned, but may also have damaging consequences for harmless species. Conversely the understatement of a hazard can result in a toll of avoidable human injuries. For most of the habitats now used for recreational purposes there is a need to re-examine the question of animal hazards in the light of modern ecological and behavioural research.

The problem of shark attacks at bathing beaches

Public attitudes towards sharks have undoubtedly been influenced by the dramatic portrayal of these creatures in recent films and novels, and also by the extensive media coverage given to incidents in which serious injuries have been inflicted on bathers. Against this background it is not surprising that the users of beach resorts in warmer regions of the world should feel apprehensive about the possibility of being attacked by a shark.

Only recently, however, has any objective attempt been made to examine the magnitude of the problem, or to relate the attacks on humans to normal shark behaviour. Much of this new information has emerged from a project initiated by the United States Office of Naval Research. This study set out to analyse eyewitness accounts of 1652 recorded shark attacks and brought together the available information in the form of an '*International Shark Attack File*' (Baldridge 1979). These findings, taken in conjunction with recent behavioural studies, refute the widespread, popular view that when sharks attack they are invariably attempting to feed on their victims. In only about 25% of the attacks documented in the File was there any evidence for this. In fact many shark-inflicted wounds appear to be produced, not as a result of biting, but by an open-mouthed slashing action using only the teeth in the upper jaw. Sharks are known to attack one another in this way in what appear to be territorial disputes. The 'bumping' of swimmers by sharks and the strange side-to-side swimming actions sometimes adopted by sharks which have been approached closely by divers (Johnson & Nelson 1973) are also likely to be forms of territorial behaviour. Unfortunately when it comes to mechanical contacts, human skin is very much more vulnerable to damage than the rough denticular covering of another shark.

One situation in which a form of feeding behaviour is undoubtedly involved is in attacks on spear-fisherman. At least twenty per cent of shark attack victims come into this category. In view of the fact that fish juices are routinely used by shark fisherman to attract shark, it is not surprising to find that the process of spearing a fish and subsequently carrying it away attached to a diving line or belt, leaking blood and body fluids into the water has a similar effect (Fig. 8.1). Sharks have been shown experimentally to detect such stimuli, which they normally associate with food, at considerable distances, certainly well in excess of 25 metres (Tester 1963). They can also sense through their lateral-line sensory system the vibrations produced by a struggling fish. Although at close quarters a shark's eyesight is certainly capable of distinguishing between a spear-fisherman and his catch (Gilbert 1963), once a shark has started to feed it seems to make little distinction between different prey items. This lack of discrimination may also help to explain the observation from the Natal coast of South

Fig. 8.1. Sharks attracted to the fish taken by a spear-fisherman are liable also to attack the spear-fisherman himself.

Africa that attacks on humans increase when sharks are feeding on shoals of fish near the shore (Wallett 1978).

Geographical distribution of shark attacks and the involvement of different shark species

Although sharks are found in all the seas of the world, records of attacks on man have a curiously limited distribution and are usually associated with tropical and subtropical water temperatures. Within the tropics (23°N to 23°S), attacks have been recorded throughout the year, whereas further from the equator (in the zone extending approximately to latitudes 40°N and 40°S) attacks have generally occurred in the local summer season (Fig. 8.2). It has been suggested (Coppleson 1962) that a temperature threshold might exist in the region of 21 °C (70 °F) which has to be exceeded for an attack to occur. On general biological grounds this seems somewhat unlikely. Marine animals in general are able to adjust their metabolic rate to the temperature of their environment, and there is no reason to believe   137

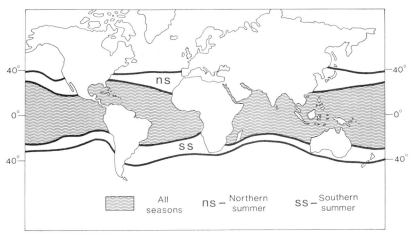

Fig. 8.2. Geographical distribution and seasonal incidence of reported shark attacks.

that sharks living in cold waters have their normal activities restricted in any way. Humans on the other hand have difficulty in remaining in water colder than 20 °C, because of an increased rate of heat loss. Bathing activity is therefore much reduced in these condition. The likely explanation for the warm water correlation is that both bathers and sharks tend to congregate in seasonally warm waters, and that it is this juxtaposition which increases the likelihood of attacks. Support for such an interpretation comes from the fact that the few recorded cold-water attacks have been on individuals who have been wearing wet suits to protect them from temperatures which would deter the average bather.

Although eyewitnesses of attacks on man are frequently unable to identify the shark species involved, tooth fragments in wounds can often be used for diagnostic purposes. From this and other information accumulated in the *Shark Attack File* it is evident that less than 30 out of a possible total of 250 shark species have been involved in attacks on man. Five of the species featuring most frequently in the list are illustrated in Fig. 8.3.

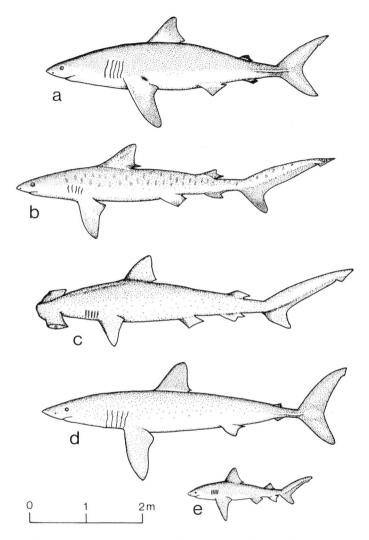

Fig. 8.3. Shark species involved in attacks on bathers in coastal areas: (a) great white shark (*Carcharodon carcharias*); (b) tiger shark (*Galeocerdo cuvieri*); (c) hammerhead shark (*Sphyrna species*); (d) mako shark (*Isurus oxyrinchus*); (e) bull shark (*Carcharhinus leucas/gangeticus*).

139

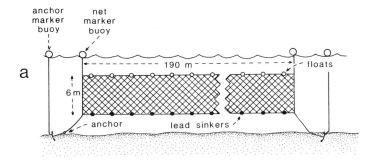

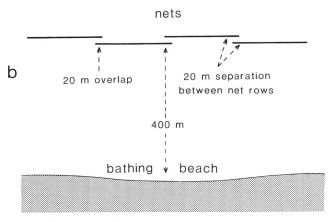

Fig. 8.4. Shark nets: (a) method of setting net; (b) typical arrangement of nets in relation to the beach (after Wallett 1978).

Control measures against sharks

In a number of warm water areas with a tradition of sea bathing, pressure by the public has led to the introduction of various measures designed to reduce the risk of shark attack.

NETS. The most widely used method of beach protection is to arrange a series of wide-meshed gill nets offshore (Fig. 8.4). The function of these nets is principally not to create a barrier, indeed the sharks can swim around the net, but to trap large sharks continuously and so reduce their numbers. The respiration of a shark depends on the flow of water over the gills produced by its forward movement, so any shark that becomes tangled in the nets dies from asphyxiation. The nets need to be inspected regularly, usually on a daily or two-daily basis, to remove dead sharks, and to repair or replace damaged nets. Shark netting was started in Australia in 1937 and in South Africa in 1952. It usually has the effect of dramatically reducing the populations of large sharks in the vicinity of netted beaches. In some localities, such as the Sydney beaches in Australia, it has had the desired effect of bringing shark attacks to a halt (Coppleson 1950, 1962). This is not always the case, however, as has been demonstrated at Amanzimtoti beach in Natal where six attacks have occurred in spite of the presence of nets (Wallett 1978).

SIDE EFFECTS OF NETTING. It has become increasingly apparent in recent years that the practice of netting has ramifications which extend beyond its effects on sharks. Along parts of the Queensland coast in Australia it has been responsible for a dramatic reduction of the local population of dugongs (*Dugong dugong*). The dugong is an interesting marine mammal, closely related to the manatee (p. 14). It feeds on marine plants and survives somewhat precariously in various coastal areas around the Pacific Ocean. Since dugongs are mammals and air-breathers they drown very quickly when caught in shark nets. Heinsohn (1972) has shown that in the first year of shark-netting operations around Magnetic Island in Queensland 82 dugongs were caught, and this probably represented the bulk of the local population. If repeated elsewhere, losses of this magnitude could have serious consequences for the survival of the species. Dugongs are not the only harmless species that become trapped in the nets. Paterson (1979) has shown that during the period 1962–78 the tally of harmless species killed in shark nets along the Queensland coast included in addition to 468 dugongs, 317 dolphins and 2654 turtles.

Another area of concern is that the removal of large predators such as sharks from a biological community may seriously disrupt its structure. The consequences of this can be as unwelcome to angling interests as the shark hazard is to bathers. Over the last 20 years sport fisherman along the Natal coast have become increasingly concerned about the decline in catches of bony (teleost) fish (Fig. 8.5d). This has coincided with an increase in the number of shark nets in operation and a decline in the numbers of large sharks (Fig. 8.5a, b). 141

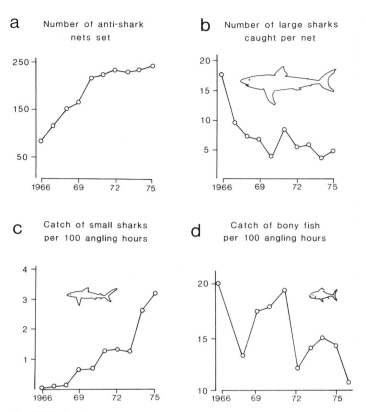

Fig. 8.5. Fisheries data suggesting relationships between: (a, b) the reduction of large sharks caused by netting; (c) the increased numbers of small sharks; (d) the decline of bony fish populations of interest to anglers (after Van der Elst 1979).

The suggested connection between these events is that the removal of the large sharks by netting allowed proliferation of the small shark species (Fig. 8.5c) which had previously been suppressed by predation from the larger species. The numerous small sharks (particularly the dusky shark) may then have been responsible for overexploiting and depressing the stocks of bony fish (Van der Elst 1979). If this interpretation of events is correct, then before netting, the large sharks were acting in the role of 'keystone species' (Paine 1966, 1980) and their removal would be expected to destabilise the community.

An alternative method of reducing shark populations would be to devise a system for catching them continuously on baited lines, a technique which has been used for many years by commercial shark fishermen. Whilst this would avoid adverse effects on harmless herbivores such as dugongs, it would not necessarily avoid the disruption of biological communities resulting from the removal of key predators.

THE USE OF BARRIERS. A quite different protective strategy and one which avoids the wholesale slaughter of sharks is to create a physical barrier between sharks and bathers. From time to time rigid meshwork enclosures have been constructed for this purpose on a number of Australian and South African beaches. These stuctures have the disadvantage of being unsightly and very difficult to maintain; the metal parts corrode and the whole structure is liable to be damaged by the surf. Not surprisingly therefore this type of barrier has lost favour. There is, however, the possibility of another type of barrier which involves an electrical field.

Sharks are known to be extremely sensitive to electrical fields. They are able for example to detect a small fish completely buried in the sand, by picking up the weak electrical field associated with its body surface (Kalmijn 1971, 1977). If subjected to strong electrical fields generated artificially, sharks at first show an avoidance reaction. With increasing field strength they begin to lose voluntary control of their swimming movements, often being drawn to the positive electrode used to pass the current into the water. This phenomenon of 'electrotaxis' is caused by the electrical current interfering with the transmission of nerve impulses. Experiments in South Africa have demonstrated both in the laboratory and in the sea that it is possible to induce electrotaxis in sharks with only a relatively modest expenditure of power (Smith 1974). In experiments in a laboratory tank, sharks failed to penetrate an electrical field ranging in strength from 5.5 to 10 V/m, created by passing pulses through a double row of electrodes (pulse duration 0.8 ms, pulse repetition rate 15 pulses/s). In an outdoor experiment, a field with similar electrical characteristics was created across a 30 m width of the St Lucia estuary in Natal. In this case the electrodes were laid parallel about six metres apart, and to test the effectiveness of the system a net was set in the water above. The system was activated on alternate days and the nets examined for captured sharks, or large holes in the net indicating that sharks had been caught

Fig. 8.6. The installation of an electrical shark barrier at the bathing resort of Margate on the Natal coast of South Africa: (a) the electrical cable in its protective pipe castings before being manoeuvred into position; (b) the cable supported by floats prior to being lowered to the sea bed (photographs by courtesy of Dr E. D. Smith, National Physical Reseach Laboratory, Pretoria).

but had broken free. The results showed that whereas 28 sharks had been trapped when the power was off, none had been caught when the electrodes were energised. A larger-scale electrical barrier has been installed at Margate (Fig. 8.6), a popular bathing beach on the Natal coast, to test the feasibility and cost effectiveness of this method compared with netting (Smith 1973, 1979a, b). The special sensitivity of sharks to electrical currents apparently ensures that the field strength used will have no effect on the passage of bony fish of interest to anglers. Nor will it have any effects on man (except conceivably in the unlikely event that five swimmers chose to hold hands in a chain directly over the cable).

The justification for control measures

The analysis derived from the *Shark Attack File* coupled with our better understanding of shark biology contradicts the widely held view that these animals are habitual man-eaters. Many attacks appear to be connected with feeding and to represent some form of territorial behaviour. A proportion of the remainder seem to have been prompted by the powerful stimuli generated by spear-fishing activities. Compared with other hazards, the overall risk of being attacked by a shark is relatively small. The total number of attacks reported annually averages only 28, with a probable fatality rate of 15–20 (Baldridge 1979). The chance of being killed by a shark is therefore considerably less than the risk of being struck by lightning, or dying as the result of a wasp or bee sting, both justifiably regarded as remote possibilities.

Some planners maintain that expenditure on measures taken to combat a hazard should match the magnitude of the risk it represents. Others feel that the amount of public concern felt about a problem should also be recognised as a relevant factor. On this latter score it has to be recognised that public attitudes are sensitised, perhaps irreversibly, on the shark issue.

If protective measures are judged to be necessary, the information now available on the side-effects of netting argues in favour of less-destructive techniques. A further consideration must be the increasing evidence that sharks possess physiological mechanisms of possible medical significance to man. Heparin-like compounds found

145

in some sharks are known to have useful anticoagulant properties (Ronsivalli 1978) and recent experiments have demonstrated that extracts from the fins and vertebrae of sharks are capable of inhibiting tumour growth in experimental animals (Lee & Langer 1983). This latter finding correlates with the general absence of tumours in sharks. It would be regrettable if destructive shark-control measures ever became so comprehensive that key species became unavailable for biological studies of this kind. Claims that populations of the grey nurse shark (*Odontaspsis taurus*) are declining along sections of the Australian coast where shark-control measures are in operation (Taylor 1977) lend support to this notion.

Other bathing beach hazards – stingrays and jellyfish

The publicity that has been given to shark attacks has disguised the fact that, in some parts of the world, other marine animals are numerically more important causes of injury or even death on bathing beaches.

Stingrays

In a survey of a much-used section of the Californian coast between Santa Monica and San Diego (208 km in length) 474 people are reported to have been injured by stingrays in a single season (Russell 1953). No figures are available for the overall incidence of such injuries on a worldwide basis, but the problem is certainly not uncommon in many warmer coastal waters.

Stingrays have the habit of lying buried in the sand of bathing beaches. Some of the time they are actively feeding on small worms, molluscs and crustaceans which they excavate from the mud using their fins, the rest of the time they are simply resting. Injuries occur when bathers step on the flat body of the fish. The stingray's response is to swing its tail forwards and upwards, driving the poison spine on the tail into the bather's foot or leg (Fig. 8.7). Venom is released into the wound by the rupture of a sheath surrounding the spine. The victim usually suffers intense pain for a period of between six and forty-eight hours and may experience more generalised circulatory and respiratory upsets (Russell 1953, 1965). Fatalities from stingray

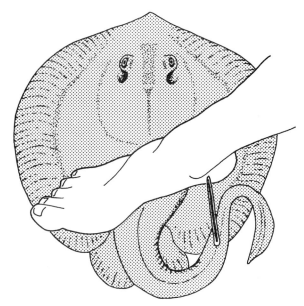

Fig. 8.7. Defensive reaction of a stingray to being stepped on by a bather (after Halstead 1978).

injuries usually come about as the result of bathers accidentally lying on the ray so that the spine penetrates the heart, abdomen or lung. As far as the stingray is concerned the venomous spine is a device for fending-off attacks by large predators such as sharks. It is unfortunate for bathers that the act of stepping on the fish elicits a similar defensive reaction.

The incursion of large numbers of stingrays into shallow waters appears to be a periodic occurrence. Some system of monitoring these movements by beach patrols, with the aim of closing beaches at critical times, would, therefore, appear to be the best approach to this particular problem.

Jellyfish

Along the north and east coasts of Australia bathing is halted periodically because of the appearance of another venomous marine animal, the box jellyfish (*Chironex fleckeri*) (Fig. 8.8b). This organism

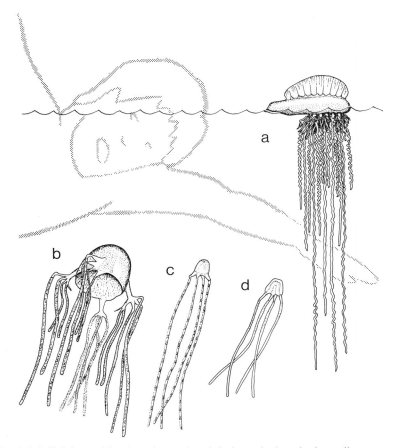

Fig. 8.8. Jellyfish capable of causing serious injuries to bathers in Australian waters: (a) Portuguese man-of-war (*Physalia physalis*); (b) box jellyfish (*Chironex fleckeri*); (c) irukandji (*Carukia barnesi*); (d) rastoni (*Carybdea rastoni*).

is known to have been responsible for at least 40 fatalities in the region (Williamson *et al.* 1980), thus considerably exceeding the estimated 27 deaths from shark attacks. Swimmers are stung when they collide with the animal which swims strongly by pulsating its bell and is not easily seen owing to its transparency. On collision, lengths of tentacle become attached to the bather's skin and toxins are injected into the body from stinging cells. The main effect of these toxins is to cause severe muscular contractions and this can have serious consequences by interfering with the normal action of the heart muscles (Endean 1981). Even if the victim survives the initial effects, the weals caused by the tentacles usually take months to heal and are likely to result in permanent scarring. The severity of the symptoms is roughly proportional to the length of tentacles coming into contact with the victim's body.

Immediate application of solutions of aluminium sulphate or dilute acetic acid can be used to reduce the discharge of stinging cells, and recently an antivenin has been developed which greatly increases the chances of recovery from severe attacks (Hartwick *et al.* 1980).

Box jellyfish appear sporadically off the Australian beaches during the southern summer season between November and March. At major beaches it has become the practice to organise beach patrols to check regularly for jellyfish and to notify holidaymakers by posting warning notices and through the press and radio. When box jellyfish are present, swimmers can be sure of avoiding injury only if they cover the entire body surface with clothing or a wetsuit.

Two other dangerous jellyfish which appear from time to time in Australian waters are *Carukia barnesi* and *Carybdea rastoni*) (Fig. 8.8c, d). The problem with these species is that there is usually a time interval of about twenty minutes between the apparently mild sting and the onset of general symptoms. Swimmers need to be more generally aware of this fact so that they can leave the water before becoming incapacitated.

Dangerous jellyfish are not confined to Australian waters and with the expansion of coastal recreation may in future create problems in other areas. The box jellyfish, for example, is widely distributed in the western Indo-Pacific region and has been responsible for fatalities in the Philippines and in East Malaysia (Williamson *et al.* 1980). The Portuguese man-of-war *Physalia* (Fig. 8.8a) is occasionally blown into temperate waters, but is essentially a tropical species. It differs from typical jellyfish in being a colonial organism in which long stinging-tentacles are attached to a gas-filled float. The float functions as a sail and allows the colony to be dispersed by the wind. Contact with the tentacles produces a characteristic weal resembling a string of beads, which usually fades after 24 hours. However, sometimes more generalised effects are produced with fatal consequences (Halstead 1978).

Biological hazards associated with coral reefs

The last decade has seen a marked expansion in the use of coral reefs for recreational purposes. Spectacular coral formations and an abundance of colourful fish and other marine animals make these especially rewarding habitats to explore. This can be done by walking on the reef at low tide or by swimming over it using a snorkel or scubadiving equipment.

Some of the species likely to be encountered are potentially dangerous because they possess chemical defence mechanisms similar to those described earlier for stingrays and jellyfish. The lion fish (*Pterois volitans*) (Fig. 8.9a) is a conspicuous red and white species which swims boldly around the reef and may actively approach swimmers. If they attempt to touch it, the fish is likely to roll its body forward and administer a sharp jab with the long poison-spines on its back. These are sufficiently sharp to penetrate diving gloves and the wounds inflicted cause long-lasting pain and, sometimes, serious generalised symptoms (Halstead 1978).

Surgeonfishes (*Acanthurus* species) represent another common group of reef fishes that swimmers need to treat with respect. These animals are equipped with knifelike spines which when extended on either side of the tail, can be used to inflict deep slash wounds (Fig. 8.9b). Venom is injected at the same time and the injuries cause intense pain. They do not, however, have the potentially serious consequences of the wounds inflicted by lion fish. Surgeonfish are usually stimulated to attack when apparently cornered by a swimmer.

Shell collectors need to be aware of the venom equipment possessed by some cone shells (*Conus* species) (Fig. 8.9d). This takes the form of a venom-laden dart which the animal extrudes from its proboscis when immobilising other small organisms for food. If the shell is picked up by a collector the dart is likely to be driven into the hand. The venom from some species can cause respiratory failure (Kohn *et al.* 1960, McMichael 1971).

The other group of problems arises in connection with venomous species which lie inconspicuously on the reef floor and are likely to be trodden on accidentally. The stonefish (*Synanceia verrucosa*) (Fig. 8.9c) is one of the more notorious of these. It has a warty body surface and is well adapted to remain undetected amongst coral rubble

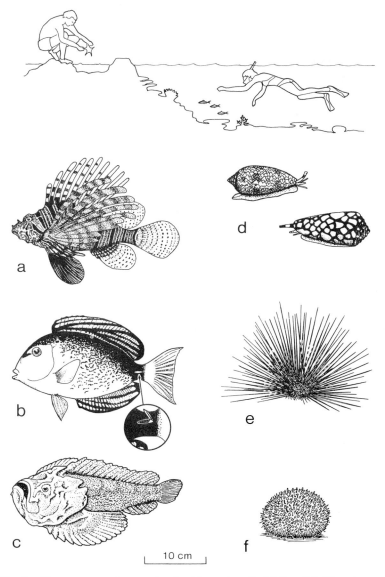

Fig. 8.9. Venomous species associated with coral reefs: (a) lion fish (*Pterois volitans*); (b) surgeon fish (*Acanthurus xanthopterus*) (inset shows extended tail spine); (c) stonefish (*Synanceia verrucosa*); (d) cone shells (*Conus textile, Conus marmoreus*); (e) needle-spined sea–urchin (*Diadema setosum*); (f) flower urchin (*Toxopneustes pileolus*).

151

where it lies in wait for shrimps and small fish. Along the back, and partially incorporated into the dorsal fin, is a series of 13 poisonous spines which it normally uses to protect itself from attacks by larger predatory fish. The spines can equally well penetrate the foot of an inadequately shod diver or reef walker, and cause serious poisoning (Cameron 1981).

Sea-urchins represent another important group of reef organisms possessing dangerous spines. The needle-spined sea-urchins (*Diadema* species) (Fig. 8.9e) have spines up to 25 cm long which readily penetrate rubber-soled footwear. The spines cause mechanical damage when they penetrate the skin and probably also inject a toxin. Certainly the short spines and pedicellariae (small pincer-like protective structures) of the flower-urchins (*Toxopneustes* species) (Fig. 8.9f) are known to incorporate venom sacs and contact with these structures can cause generalised symptoms including respiratory paralysis (Edmonds 1975).

Strategies for combating venomous marine organisms

The view that organisms which pose a hazard in recreational areas should simply be removed or eliminated, has little to commend it in ecological terms. In many situations such actions are likely to lead to the serious disruption of biological communities (Paine 1966, 1980). For example, the experimental removal of needle-spined sea-urchins from reefs in the West Indies has been shown to stimulate excessive growth of sea 'grasses' (Ogden *et al.* 1973). The urchins normally keep these plants in check by means of their grazing activities. Similar disturbances could be anticipated from the removal of key carnivorous species. This is a particularly relevant consideration in the case of coral reefs whose attraction for visitors lies largely in the opportunities afforded to see species interacting normally in their natural habitat.

A better approach is for visitors to be taught to recognise the more conspicuous of the dangerous species and to be given advice about ways of protecting themselves from the cryptic ones. Stonefish and sea-urchins come into this latter category and penetration by their spines can usually be prevented if reef-walkers use leather-soled footwear and swimmers wear flippers with full heel protection. Where a species occurs only periodically, as is the case with stingrays and jellyfish on bathing beaches, the appropriate action is to monitor the situation and close beaches to unprotected swimmers when the animals appear.

Man happens to be a thin-skinned and physiologically sensitive animal who has chosen to invade, for his own amusement, a variety of marine habitats already occupied by species which use powerful chemical devices to carry out their everyday business. In these circumstances it is not unreasonable to expect man to accept some constraints on his leisure activities.

Land and inland-water hazards

Travellers' accounts of attacks by large and ferocious animals have an even longer history in inland habitats than at the coast. The species mentioned range from grizzly bears and wolves, to giant snakes and crocodiles. Our modern knowledge allows us to reject at least a proportion of these accounts as fanciful. From recent studies on wolf behaviour it seems very unlikely that these animals ever habitually harassed human travellers (Mech 1970, Lopez 1978). Similarly few of the reports of attacks on man by giant constricting snakes such as anacondas, pythons and boa constrictors survive close examination (Pope 1962).

There remains, however, a residual list of species whose reputation for attacks on man is borne out by modern observations. Our present purpose is to examine such occurrences in the context of tourism and recreation. The most appropriate groups to consider are crocodiles. poisonous snakes and bears.

Crocodiles

Of the 25 species of crocodiles only two are confirmed as making regular unprovoked attacks on man (Schmidt & Inger 1957). These are the Nile crocodile (*Crocodylus niloticus*) from Africa and the saltwater crocodile (*Crocodylus porosus*) from South East Asia and Australia. In some rivers and lakes in Africa the Nile crocodile has a long history of attacking humans, apparently for the purpose of feeding on them. Local people swimming, wading, falling from boats, trailing their hands in the water, or filling water pots at the river's

Fig. 8.10. Sign in the Kakadu National park in Northern Australia warning
visitors of the hazard from saltwater crocodiles.

edge are all potentially at risk. The threat from the saltwater crocodile
occurs mainly in the brackish waters of river estuaries (Webb *et al.*
1978, Kar & Bustard 1983). To what extent these animals might
represent a hazard to tourists has yet to be fully explored. It is
interesting to note, however, that in at least one recreational area, the
Kakadu National Park in Northern Australia, an energetic publicity
campaign has been mounted to deter visitors from swimming in areas
frequented by saltwater crocodiles (Fig. 8.10).

Hazards from poisonous snakes

Snakes bite an estimated half a million people every year and of these
between 30 000 and 40 000 die (Swaroop & Grab 1954). Without
question therefore they represent the most significant of all the larger
animal hazards.

Certainly, prospective visitors to the tropics often feel apprehensive
about snakes and superficially the snakebite statistics from different

Table 8.1. *Mortality rates from snakebite in different countries, expressed as deaths per 100000 of population per annum[a]*

India	5.40	Japan	0.13
Sri Lanka	5.10[b]	Australia	0.07
Venezuela	3.10	France	0.06[d]
Burma	3.30[c]	Italy	0.04[d]
Costa Rica	1.93[d]	Spain	0.02[d]
Colombia	1.56[d]	England & Wales	0.02[d]
Thailand	1.30[d]	Canada	0.02[d]
Mexico	0.94	USA	0.01[e]

a Data from Swaroop & Grab (1954) unless otherwise indicated.
b De Silva (1980).
c Aung-Khin (1980).
d Includes bites and stings of all venomous animals.
e Parrish (1959).

countries appear to support their view. Mortality from snakebite is generally far greater in tropical regions than in temperate ones (Table 8.1). From the tourist's standpoint, however, these national statistics are liable to be misleading. Where more detailed studies have been made of the circumstances of snake attacks in the tropics, it becomes apparent that it is the local population which is at risk rather than the visitor. In a fairly typical case study from the State of Anuradhapura in Sri Lanka, a region which has one of the highest snakebite mortality rates in the world, virtually all the incidents involved local people in rural communities (De Silva 1981). This was in spite of the fact that Anuradhapura, because of its historic interest as Sri Lanka's ancient capital, is a major tourist centre. Outdoor attacks were typically directed against people cutting grass or weeding crops. The species involved in these circumstances were usually cobras (*Naja naja*) or Russell's vipers (*Vipera russelli*). The vulnerability of local people was clearly increased by the frequent lack of protective footwear. Indoor attacks were mainly a result of kraits (*Bungarus caeruleus*) entering houses at night in search of rodents and attacking people sleeping on the floor.

In these circumstances it is not surprising that tourists, with their access to superior living accommodation, their better footwear and

a

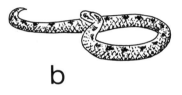

b

c

25cm

Fig. 8.11. Venomous snakes likely to be associated with African 'safari' camps: (a) puff adder (*Bitis arietans*); (b) night adder (*Causus maculatus*); (c) black-necked cobra (*Naja nigricollis*) (after Cansdale 1961 and Stidworthy 1974).

156

their general lack of contact with snake-infested vegetation, generally avoid attacks.

Whether this is invariably the case in tropical environments is not clear. In some African 'safari' camps the facilities created for tourists have features in common with local villages, in fact this resemblance may be actively fostered to provide an authentic atmosphere. Recent work on snakebite in African villages has shown that night adders (*Causus maculatus*) and puff adders (*Bitis arientans*) (Fig. 8.11a, b) are a major problem in the vicinity of houses at night (Warrell *et al.* 1975, 1976*a*, Coetzer & Tilbury 1982). The puff adder emerges from its hiding place to lie in wait for rats on bare sandy areas where its camouflage makes it difficult to see. If the snake is disturbed it strikes out at the foot or calf and the injuries caused can lead to circulatory collapse and death. The smaller night adder behaves in a similar way when hunting for frogs and toads. Bites on humans from this species usually involve the toes, foot or ankle but rarely result in serious poisoning. The third problem-species is the black-necked cobra (*Naja nigricollis*) (Fig. 8.11c). Although this species is best known for its ability to defend itself by spitting venom at the eyes of intruders, it has recently become apparent that it will also enter houses at night in search of frogs, rats and poultry, and if disturbed, will strike at humans (Warrell *et al.* 1976*b*).

Most safari camp huts are presumably sufficiently snake-proof to prevent the entry of cobras. Night adders and puff adders may, however, represent a more significant hazard, because of the tendency of visitors to move about in the camp after dark, either in connection with animal-watching activities or when walking between sleeping quarters and centralised camp facilities. Additionally, prey species attractive to snakes, such as frogs and rats, are not uncommon at such sites; indeed rat populations are often favoured by inadequate refuse disposal arrangements. In such circumstances the possibility of a snakebite problem developing deserves attention.

Snakebite hazards in temperate regions

In spite of the lower general incidence of snakebite in temperate countries, there are reasons for believing that the proportion of attacks specifically associated with recreation is actually higher than in the

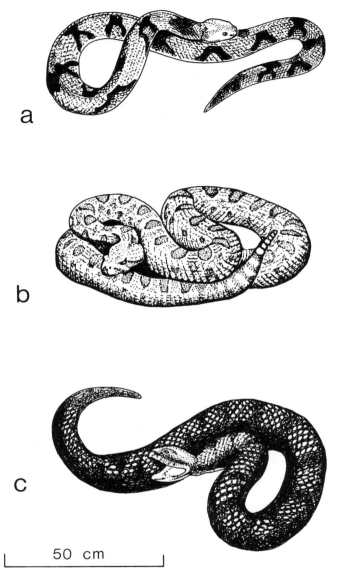

Fig. 8.12. Snakes most frequently involved in attacks on man in the United States: (a) copperhead (*Agkistrodon contortrix*); (b) rattlesnake (*Crotalus species*); (c) cottonmouth (*Agkistrodon piscivorus*) (c after Stidworthy 1974).

tropics. An analysis of snakebites in the United States showed that 15% of all bites were received by children involved in play activities of some kind, and that a further 8% were associated with adult recreational pursuits such as hunting and fishing (Parrish *et al.* 1965). In 31% of cases the attacking species was a copperhead (*Agkistrodon contortrix*), in 29% of cases a rattlesnake (*Crotalus* species) and in 9% of cases a cottonmouth (*Agkistrodon piscivorus*) (Fig. 8.12). In describing the circumstances of bites caused by rattlesnakes, Klauber (1972) similarly stresses the vulnerability of people involved in leisure activities. He describes how hunters are liable to be bitten when stalking game or when attempting to retrieve shot animals from holes or from vegetation. Bites on fishermen tend to occur when they disturb rattlesnakes lying amongst rocks on river banks.

Preventive and remedial measures against snakebite

An obvious first approach to avoiding snakebite is to learn something about the places where snakes are most likely to be found and consciously avoid putting unprotected hands and feet into these situations. The appropriate precautions differ from one region to another but in most snake-frequented areas the dangerous places include thick vegetation, hollow trees, animal burrows, logs, and bare-ground surfaces at night.

If a bite is sustained, the chances of successful recovery are greatly enhanced by admission to hospital for observation and if appropriate, treatment with an antivenin. It is important to counter the widespread belief that extremely unpleasant and often fatal consequences necessarily result from a snakebite. When striking defensively at man, venomous snakes commonly use less venom than when attacking smaller animals which they intend to use as food. In fact, about 30% of strikes on man by viperid snakes (vipers, adders and rattlesnakes) involve insufficient venom to produce symptoms of poisoning. The comparable figure for elapid snakes (e.g. cobras, kraits, mambas) is 50%. An appreciation of this fact might help to counter the serious fright and shock reactions frequently associated with snakebite, even when no venom has been injected. The venoms produced by viperid snakes consist of vasculotoxins which act on the blood and circulatory system. With these snakes the absence of swelling at the bite site in the first half hour is usually a reliable indication that no poison has been injected. With elapid snakes, where a neurotoxin (affecting nerves and muscles) is typically involved, a longer period of some hours may need to elapse before it becomes apparent whether or not serious poisoning has occurred. However, because of the difficulties frequently encountered in determining the identity of the attacking species, admission to hospital for observation is a wise precaution in all cases (Reid 1983, Reid & Theakston 1983).

Possibilities for reducing the incidence of snakes by habitat modification are rather limited. The removal of cover such as logs, boards, rocks and refuse, together with the blocking of rodent holes is reported to reduce the incidence of rattlesnakes in the vicinity of homesteads, and this is an approach which is equally applicable to small recreational areas. Fences can be used to exclude rattlesnakes from facilities such as childrens' playgrounds. However, the fences need to be at least two metres high and to consist of fine mesh sunk below ground level (Klauber 1972). In Japan, low (65 cm) electric fences have been used in an attempt to exclude a troublesome snake known as the habu (*Trimeresurus flavoviridis*) from villages (Hayashi *et al.* 1979). This has proved only partially successful because of the difficulty of providing human-access gates which are snake-proof.

The more extreme strategy of attempting to destroy snakes on a large scale is only really applicable to North temperate species such as certain rattlesnakes which overwinter communally in underground dens (Klauber 1972). Even then, there are considerable ecological objections to such measures. Harmless species which occupy the same dens are likely to be killed in the process, and the general removal of snakes could lead to increases in troublesome rodent populations. In national parks it is axiomatic that such disruptions of natural communities should be avoided whenever possible.

The North American black bear – an underestimated hazard

In contrast to the general tendency to overestimate large animal hazards, in at least one case the converse has been true. For many years the capacity of the North American black bear (*Ursus americanus*) to create difficulties in recreational areas has been seriously underestimated.

Fig. 8.13. Black bear begging for food from a visitors' car in the Jasper National Park (photograph – Baribal/Okapia).

Throughout the initial development of the parks in the 1920s and 1930s the prevailing image of the bears was of friendly, playful animals, capable of being coaxed into closer proximity with titbits of food. During this phase, visitors were permitted to feed bears at campsites and along the roadside (Fig. 8.13). In the Yellowstone Park, for example, it was not uncommon to have as many as a dozen bears begging for food along a 10-mile stretch of road (Brown 1982). Both at Yellowstone and in the Yosemite Park this roadside feeding of bears was a common cause of traffic jams. Neither were bears discouraged from scavenging for food at open refuse tips. Indeed this habit was turned to advantage by the park authorities as a means of displaying the animals to the public. At both Yosemite and Yellostone seating was arranged around refuse tips to allow visitors to enjoy the spectacle (Graber & White 1978, Brown 1982). In the Sequoia National Park, rangers conducted a regular schedule of bear feeding for the same purpose.

HABITUATION TO HUMAN FOODSTUFFS. Since this period it has become abundantly clear that encouraging such traits in bear populations was not in the best long-term interests of either bears or visitors. Bears are well known to be opportunistic feeders, and it was not surprising that they should have learned quickly to exploit the energy-rich materials represented by human foodstuffs and food-wastes. What was not appreciated initially was the extent to which this habit would transform the behaviour of the animals.

Black bears emerge from hibernation in the spring and initially feed on grasses and other vegetation until this material becomes dry and less succulent. In autumn they concentrate on nuts and berries. In midsummer, however, they are obliged to feed in an opportunistic manner on any food sources they can find. Sometimes these sources take the form of crickets and grasshoppers, sometimes attention is turned to the nests of social insects such as wasps, bees and termites. The opportunity to utilise human foodstuffs during this potentially lean midsummer period was therefore one which was readily accepted and incorporated into the animals' regular annual cycle. This dependence, coupled with the public's tendency to underestimate the strength and aggressiveness of bears, led to an increasing toll of human injuries and damage to property. Bears developed the habit of searching aggressively for food in campsites, tearing open tents and

Fig. 8.14. Bear-proof refuse hopper in the Yosemite National Park.

sometimes breaking into vehicles and cabins. Inevitably this led to confrontations involving human injuries. In one season in the 1930s at Yosemite more than 60 people required hospital treatment as a result of injuries caused by bears, and similar problems were being experienced in the Yellowstone Park.

REMEDIAL ACTIONS. The initial response of park authorities to the growing number of incidents involving injury and damage to property was to move bears from areas where they were causing problems, or to kill them. Bears can be moved by tranquilising the animals with a narcotic dart, and transporting them to new localities by means of trucks or helicopters. Predictably, relocation can have the effect simply of transferring the problem to another area, whilst in many instances troublesome bears have found their way back to their original starting points. Moreover there are increasing indications that the immobilising drugs may themselves produce unwelcome behavioural changes. The extent to which bears have been intentionally killed in control operations is poorly documented for many parks. In the Yellowstone Park, however, control actions in the 1950s are reported to have involved killing about 30 bears a year (Brown 1982).

An alternative approach, and the one increasingly being adopted,   163

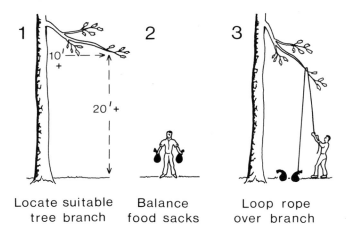

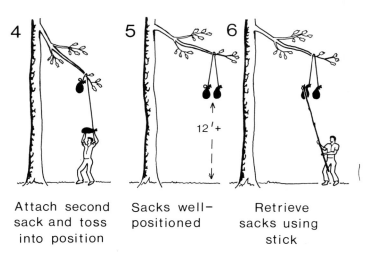

Fig. 8.15. Instructions for food storage designed to prevent access by bears in back country areas of the Yosemite National Park.

is to develop ways of denying bears access to human foodstuffs. By this means it is hoped to return the bears to their normal feeding regimes and minimise confrontations with man. The technological part of this strategy has involved the design of a variety of bear-proof containers in the form of litter bins, refuse hoppers and food lockers

(Fig. 8.14). It has also included new arrangements for removing refuse from parks for disposal elsewhere. In remote areas, campers are advised to store food out of the reach of bears by hanging counterbalanced food sacks on tree branches which are too slender to support the weight of a bear (Fig. 8.15). At some sites cables have been strung between trees for the same purpose. Not all these devices have proved effective. Some animals have learnt how to operate a system which employs pulleys to lower the cable to the ground. At Yosemite a bear has been observed to shake the cable to detach poorly secured food sacks (Graber & White 1978).

Ultimately such measures are only as effective as the readiness of the public to use them. Various techniques are used to impress on visitors the importance of denying access to human foodstuffs. Posters, leaflets, children's story books and rangers' talks are all utilised for this purpose. Some powers of enforcement are also available in the form of fines. An undoubted impediment to the success of this programme is the continuing 'teddy bear' image of these animals, which has probably been re-enforced by Yogi Bear comic strips and Smokey-the-Bear fire prevention programmes.

GRIZZLY BEARS. A final point to be made about bears is that the other species in North America, the grizzly bear (*Ursus arctos*), presents a rather different combination of problems. This is a larger and more powerful animal than the black bear and few park visitors have any illusions about its tameness. Although some attacks on humans seem to have been associated with habituation to human foodstuffs (Herrero 1976), others in backcountry areas appear to have no such associations (Gildart 1981). These backcountry incidents usually occur when a hiker surprises a female accompanied by cubs, or disturbs a bear actively feeding on animal carrion. The injuries and occasional fatalities caused by grizzly bears in North American national parks have prompted some observers to press for the elimination of these animals from parks which are heavily used by the public (Moment 1968). The contrary view stresses the low rate of grizzly bear-induced injuries compared with other recreational hazards, and emphasises the need to avoid measures which might further jeopardise the survival of this endangered species (Craighead & Craighead 1971, Chase 1983).

Myth and reality in the management of large animals

Whatever the public's view of large and potentially dangerous animals, managers of recreational areas cannot afford to base their actions on the stereotypes perpetuated by nursery stories, cartoon strips and horror films. As the case of the black bear demonstrates, underestimating a hazard can lead to the development of relationships which are not in the best interests of either animals or man. Equally, the more usual tendency to overdramatise hazards can lead to the introduction of unnecessarily severe countermeasures which disrupt natural systems and put additional pressure on endangered species. In the long term such policies deprive mankind of the many practical and psychological benefits to be gained from a properly balanced relationship with the natural world.

9.
ENVIRONMENTAL EFFECTS OF TOURIST SUPPORT FACILITIES

The effects of tourists on the environment are not limited to the occasions when they are taking part in some specific recreational activity. Tourists staying for any length of time in an area require support facilities in the form of residential accomodation, roads and car parks, water supplies and waste–disposal facilities. Any review of the impacts of tourism would be incomplete without some consideration of the environmental consequences of providing these facilities.

The ecological effects of roads

Roads as barriers

The construction of a road can affect wild species in a variety of ways. Observations from the Kruger National Park in South Africa have shown that roads on embankments can act as barriers to young animals (Pienaar 1968). When family groups of ostriches, warthogs or elephants attempt to cross such embankments the young may be left behind especially if the group is disturbed. Tortoises of all ages attempting to climb the slopes are liable to roll down to the bottom and if they become overturned, inevitably die.

Even roads without embankments represent a major barrier for some species. In Germany, Mader (1984) has shown that some of the smaller forest-dwelling species such as the yellow-necked mouse (*Apodemus flavicollis*) are reluctant to cross roads even though they

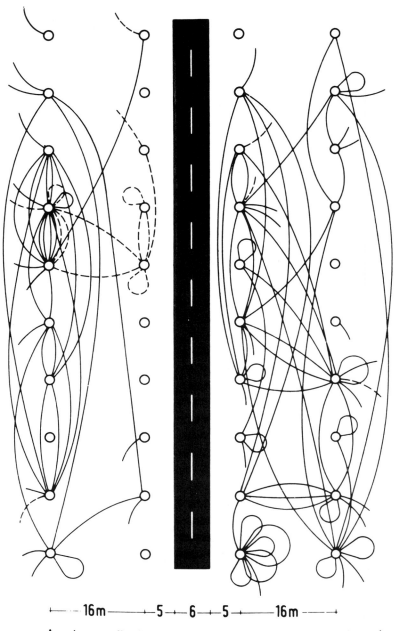

—— Apodemus flavicollis   ——— Clethrionomys glareolus

regularly travel comparable distances in other directions (Fig. 9.1). This inhibition was found to apply not only to main highways but also to a lesser extent to minor roads and even forest roads closed to the general public. Instead of the 28 crossings which would have been predicted on the basis of the animals' normal movements there were only four crossings of the minor road and two of the forest road. Similar results have been obtained with a number of beetle species normally occupying forest habitats (Mader 1984).

Opinions are divided as to the biological implications of this road-avoidance behaviour. It has been suggested that it restricts the exchange of genetic material within the animal population concerned and would therefore limit its capacity to adapt to environmental change. However, whilst this might be an important consideration for endangered or highly localised species it probably has little general significance.

Road mortality

More obvious risks face species with no inhibitions about approaching or crossing roads. These face the real hazard of being killed or injured by passing vehicles. In the Kruger Park night-driving vehicles frequently kill scrub hares (*Lepus saxatilis*) feeding on the short grasses which flourish in the moist furrows at the road edge (Pienaar 1968). In North America a high mortality rate amongst white-tailed deer (*Odocoileus virginianus*) is similarly linked with the animals' habit of feeding on the grassy margins of roads (Bellis & Graves 1971). Species of scavenging birds and mammals which make a habit of feeding on road casualties are themselves likely to become victims.

Particular problems arise when roads are constructed across animal migration routes. In Europe, naturalists have drawn attention to the fact that badgers (*Meles meles*) are likely to become road casualties when their traditional routes between setts and night-time feeding

---

Fig. 9.1. Movements of yellow-necked mice (*Apodemus flavicollis*) and bank voles (*Clethrionomys glareolus*) between live traps in the vicinity of a road. Although the animals frequently moved comparable distances in other directions they were never observed to cross the road. (Reproduced with permission from 'Animal habitat isolation by roads and agricultural fields' H. J. Mader, *Biological Conservation* **29**, 81–96. Elseviér Applied Science Publishers.)

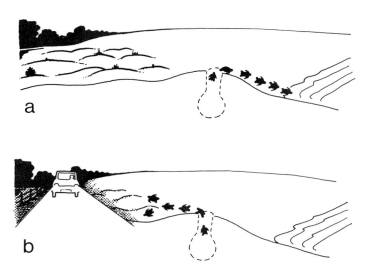

Fig. 9.2. Disorientation of sea turtle hatchlings by car headlights.

areas are bisected by new roads. Similar hazards are faced by toads (*Bufo bufo*) when they make massed migrations from winter hibernation sites to breeding ponds (Ratcliffe 1983).

Disorientation of marine turtles

A less obvious road-related problem arises in relation to the breeding activities of marine turtles. As must now be widely known from television films, many species of marine turtles lay their eggs in laboriously excavated nests situated above the tide line on sandy beaches. After about two months of development, the young turtles hatch from eggs, emerge on to the sand surface and make their way down to the sea (Fig. 9.2a). During this process they suffer attacks from a variety of predators. The relevance of roads to this pattern of behaviour is that the headlights of cars on beach-top roads can disorientate the hatchlings and cause them to crawl inland rather than towards the sea (Fig. 9.2b). Apparently young turtles normally use the brighter seaward horizon as a cue to indicate the direction of the sea (Ehrenfield & Carr 1967). Artificial lights inland disrupt this mechanism and can result in an abortive and usually disastrous mass

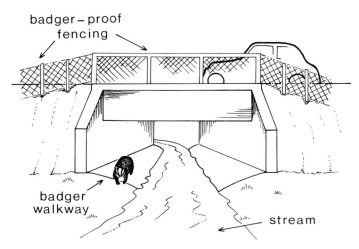

badger – proof
fencing

badger
walkway

stream

Fig. 9.3. A 'badger underpass' consisting of walkways incorporated into a stream culvert.

migration of hatchlings away from the sea. As well as car headlights, lights in beachside hotels, street lights and the floodlights of sports centres can all have this effect (McFarlane 1963, Philibosian 1976).

Remedial measures

With sufficient forethought at least a proportion of the problems created by roads can be alleviated. On turtle beaches, it has been suggested that shrubs and trees could be used to screen the beach from artificial illumination (Siow & Moll 1982). However, the effectiveness of this method has yet to be tested.

Culverts are routinely required to allow streams to pass underneath roads, and it is often not a difficult matter to modify these structures to allow the safe passage of animals. The first 'badger underpass' was constructed in 1971 on the M53 motorway on the Wirral Peninsula in Cheshire. It involved adding walkways alongside the stream in the culvert (Fig. 9.3) and the construction of a system of badger-proof fencing to guide the animals to the underpass (Ratcliffe 1983). Subsequent designs have involved channelling the animals through rounded conduit pipes. Underpasses for deer need to be much larger

171

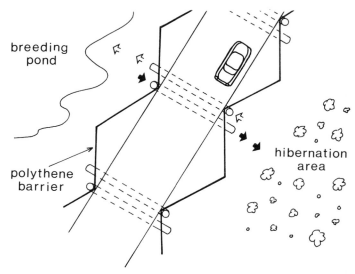

Fig. 9.4. Under-road tunnels designed to facilitate the movement of toads between hibernation areas and breeding ponds (British Herpetological Society 1984).

and to have an overhead clearance of at least two and a half metres. In addition, guidance fences for deer need to take account of their jumping capabilities and be at least two metres high.

In several European countries underpasses have also been constructed for toads. The animals are guided by 50-cm-high polythene fences to the entrance holes of sloping tunnels passing under the road (Fig. 9.4) (British Herpetological Society 1984). Other ways of assisting toads involve temporary road closures, the erection of warning signs, and the transfer across the road in buckets, of toads which congregate at specially erected trapping fences (Leeuwen 1982).

Most of these initiatives have been taken in Western countries, and some of them depend on a committed, volunteer, labour force. However, the idea of making minor constructional alterations to roadways, with the aim of reducing animal mortality, is universally applicable and deserves to be given wider attention, especially in National Parks.

Environmental problems associated with waste-disposal practices at tourist resorts

A quite different set of problems can arise if insufficient attention is given to the disposal of organic wastes from tourist areas. The main contribution of ecology in this context is to explain how species which utilise such materials can generate serious management problems.

Excessive algal growths in inland waters enriched by sewage effluents

The most significant nuisance involving plants comes from the excessive growth of algae which develop in recreational waters enriched by sewage effluents. The most commonly used means of sewage treatment, involving primary and secondary stages, is not sufficient to remove nutrients which stimulate plant growth from the final effluents. These effluents may contain as much as 8 mg/litre of phosphate-phosphorus and 20 mg/litre of inorganic nitrogen. The comparable values for unpolluted lake water are of the order of 0.01 and 0.10 mg/litre respectively. Nutrient-rich discharges into enclosed water bodies can stimulate algal growth to such an extent as to seriously disrupt recreational activities.

Sometimes the offending organisms take the form of filamentous algae (Fig. 9.5a) the strands of which merge together to form dense floating masses at the water surface. These so-called 'blanket weeds' can interfere with angling by becoming entangled with fishing lines. They can also make swimming extremely unpleasant. Another kind of troublesome growth consists of dense concentrations of individual algal colonies. Species such as *Microcystis* and *Anabaena* (Fig. 9.5b, c) can appear in such profusion as to make the water of a lake cloudy, a phenomenon often referred to as a 'water-bloom' (Brook 1957). This condition can make swimming unpleasant and in sensitive individuals is likely to cause skin rashes and gastrointestinal upsets. Angling can also be adversely affected because of the release by the algae of cell products which are toxic to fish (Gorham 1964).

Finally both the blanket-weed algae and the bloom-producers can create an additional nuisance if they are washed up on lake shores. Here they form evil-smelling masses and may provide a breeding ground for biting stable flies (p. 130).

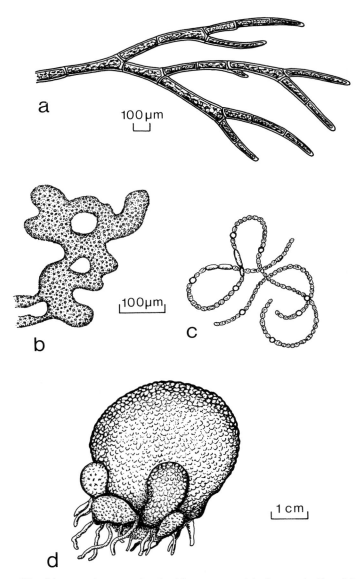

Fig. 9.5. Troublesome algae associated with sewage-enriched water bodies: (a) the blanket weed, *Cladophora*; (b, c) two bloom-forming, blue-green algae, *Microcystis* and *Anabaena*; (d) the 'green bubble alga', *Dictyosphaeria cavernosa*. (Based on Palmer 1962, Belcher & Swale 1976.)

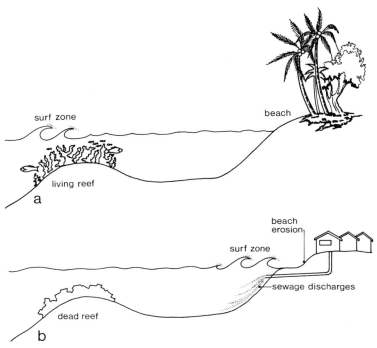

Fig. 9.6. The destruction of coral reefs by excessive algal growths not only destroys the recreational interest of these habitats, it may also lead to accelerated beach erosion (after Odum 1976).

Sewage discharge and coral reefs

Similar problems associated with sewage effluents have arisen on some coral reefs, particularly reefs in bays where tidal movements are limited. On the Hawaiian island of Oahu the discharge of partially treated sewage effluents into Kaneohe Bay has stimulated the growth of the alga *Dictyosphaeria cavernosa* to such an extent that it has overgrown and killed large sections of the reef (Johannes 1971, 1975). The alga first forms spherical colonies (Fig. 9.5d); these then spread and coalesce to form an almost continuous blanket which covers the reef. Corals are filter-feeding animals whose nutrition depends partly on suspended organic particles obtained from the water, and partly on carbon compounds derived from small plant cells (zooxanthellae) which live in the animals' tissues (Muscatine & Cernichiari 1969). The

175

blanket of *Dictyosphaeria* is therefore doubly harmful, depriving the coral organisms of their suspended food particles, and also shutting off light to their zooxanthellae. In these circumstances, it is not surprising that the reef succumbs. Soil erosion caused by building activity can also be damaging to coral reefs, because it increases water turbidity which in turn reduces light penetration (Odum 1976).

The degeneration of a coral reef at a tourist resort can have many ramifications. Not only does it deny the visitor the opportunity to explore a unique type of biological community, it may also increase the vulnerability of the beach to storm damage. Fringing coral reefs have rightly been described as self-repairing breakwaters, capable of dissipating wave forces equivalent to many thousands of horsepower. If they are destroyed these same forces are directed at the beach itself, often sweeping away sand and sometimes causing damage to property (Fig. 9.6).

Prevention of sewage-related problems

It is technically feasible to reduce the concentrations of plant nutrients in sewage effluents below the level at which they are likely to cause algal problems. This involves a third, or tertiary, stage being added to the conventional treatment process. In this arrangement there is a primary stage which involves the screening and settling out of major solids; a secondary stage which employs biological processes to decompose and remove organic materials; and the tertiary stage which uses chemical, physical or biological methods to remove nutrients from the final effluent. The main reason why this final stage is so rarely included is one of cost; arrangements for tertiary treatment can actually double the overall cost of the installation.

Using such parameters as lake depth, lake-flushing times and the level of nutrient inputs from other sources, it is now possible to predict what additional nutrient input from sewage sources is likely to cause algal problems (Vollenweider 1975, Dillon & Rigler 1975). However, the simplest guideline for tourism planners is to avoid discharging any effluents which have not received tertiary treatment into confined water bodies.

Tourist establishments close to population centres are often able to solve their waste-disposal requirements by discharging into existing municipal treatment systems. As long as these systems have sufficient spare capacity to accomodate the extra load this is an ideal solution. Difficulties are more likely to arise at isolated localities where independent arrangements for waste disposal have to be made. In these circumstances the problems are aggravated by seasonal fluctuations in visitor numbers and the difficulties of arranging for adequate plant supervision (Barnes 1973). Nonetheless quite apart from the health risks mentioned in Chapter 6 (p. 112), failure to make adequate provisions carries with it the risk of producing environmental changes directly detrimental to the enjoyment of visitors.

The use of refuse by scavenging animals

Just as the nutrients in sewage effluents can be utilised by certain aquatic algae, so also can the edible components of refuse be exploited by scavenging animals. Some of these species are sufficiently large and powerful to pose a direct physical threat. There is evidence for example that the casual dumping of refuse from tourist sites into the sea can attract potentially dangerous sharks. Mention has been made earlier (p. 162) of the problems generated by the refuse-feeding habits of black and grizzly bears in North American National Parks. Visits by polar bears (*Thalarctos maritimus*) to urban refuse tips in Northern Canada (Golden 1981) suggest that this species too, has the potential to create similar problems at recreational sites if large-scale tourist development is extended to these regions.

Flocks of scavenging gulls can create difficulties because of the risk of collisions with the light aircraft often used to service tourist sites. In Africa a similar hazard is associated with scavenging flocks of Marabou storks (Fig. 9.7).

Disease carriers represent a second category of problem species amongst the scavengers. Of these, rats are most important, particularly because of their involvement in the transmission of leptospirosis or Weil's disease. This is caused by the bacterium *Leptospira icterohaemorrhagiae* and can produce serious and often fatal jaundice in man. It is passed into the water of a river or lake in rat urine and usually enters the human body through small cuts or grazes. The danger of infection arises where rat-infested refuse tips are allowed to develop alongside recreational water bodies. In the tropics the 177

Fig. 9.7. The marabou stork (*Leptoptilos crumeniferus*), a species commonly associated with refuse tips in African National Parks (after Serle *et al.* 1977).

presence of rat-infested rubbish dumps adjacent to recreational sites can have the added disadvantage of attracting dangerous snakes, which whilst they might help to control the rats, sometimes attack man (p. 157).

Amongst the smaller disease-carriers associated with refuse tips must be included the common housefly (*Musca domestica*), the lesser housefly (*Fannia canicularis*), and certain mosquito species such as *Aedes aegypti*. Whilst the precise role of houseflies in disease transmission is a matter for debate (Greenberg 1965), *Aedes aegypti* is undoubtedly an important carrier of yellow fever and dengue (p. 101). In this case the association with refuse tips probably depends less on the organic content of the material than on the habitats for larvae created by water-filled tin cans and similar containers.

A more indirect consequence of the build-up of scavengers in areas of human activity is the effect they can have on more highly valued species. For example on the islands of the Great Barrier Reef in Australia there is concern that the food obtained by scavenging at tourist sites is so increasing the populations of silver gulls (*Larus novaehollandiae*) that their attacks on the eggs and chicks of crested terns (*Sterna bergii*) are posing a threat to this species (Great Barrier Reef Committee 1978). Such problems in tourist areas can be further aggravated by visitors disturbing breeding birds thus making their nests vulnerable to attacks by gulls (p. 37). Rats attracted to refuse tips can also represent a hazard to bird-breeding colonies.

IMPLICATIONS FOR REFUSE MANAGEMENT. Much of the experience gained by municipal authorities in attempting to solve the problems of refuse disposal is relevant to tourist sites. Where refuse is being disposed of by tipping the practice of covering successive layers of waste with an inert material such as soil is designed to check exploitation by scavengers. However, this approach is not invariably successful. Larger animals such as bears are capable of digging through a soil cover, whilst rats and gulls are sufficiently bold and agile to exploit a tip in the interval between dumping and covering. Net cages can be built to exclude gulls and stout fences constructed to exclude bears, but such erections are costly and difficult to maintain and neither will deal with the problem of rat infestations. On-site incineration of refuse is another possibility but at tourist sites it is often difficult to design a system which will cope economically with seasonal fluctuations in the amount of waste material needing to be processed. Extra fuel may be required to supplement the combustible fraction present in the refuse. In a general review of waste-disposal needs of tourist sites Christiansen (1977) came to the conclusion that the most satisfactory arrangements was to transport wastes out of tourist areas for disposal at existing municipal facilities. This strategy does however require the development of efficient systems for collecting the refuse in the first place, and then for transferring it to the vehicles required for transport to an outside disposal facility.

Tourist support facilities in perspective

Although the construction of roads and the disposal of wastes provide some striking examples of the side-effects of tourist support facilities they by no means exhaust the range of possible problems. Other provisions such as those required to obtain fuel and water supplies can be equally damaging. Thus in Nepal the felling of trees to provide fuel at tourist camps has exacerbated the already serious problem of soil erosion (Eckholm 1975, Ridgeway 1982). At coastal resorts in many parts of the world the practice of obtaining water by sinking boreholes into sand dune areas has major biological consequences because it causes a lowering of the water table. This has the effect of drying-out and destroying moist 'slack' habitats together with their characteristic floras and faunas (Ranwell 1972). Another method of obtaining water, by desalination, can have equally far-reaching effects. In this case it is marine communities which suffer damage because they are unable to tolerate the hypersaline and copper-contaminated effluents discharged from these installations (Johannes 1975).

When designing tourist complexes planners routinely give careful attention to the immediate needs of future visitors (Baud-Bovy & Lawson 1977). They now need to extend their attention to the less direct consequences of the facilities they provide. Irrespective of disturbances which can be caused to wildlife populations, lack of attention to ecological factors can lead also to environmental changes which are directly prejudicial to recreation.

REFERENCES

Addison, H. (1964). *A Treatise on Applied Hydraulics.* Chapman & Hall, London.

Akre, R. D. & Davis, H. G. (1978). Biology and pest status of venomous wasps. *Annu. Rev. Entomol.* **23**, 215–38.

Ali, S. (1960). The pink-headed duck *Rhodonessa caryophyllacea* (Latham). *Wildfowl Trust Annu. Rep.* **11**, 55–60.

Ali, A. & Mulla, M. S. (1978*a*). Declining field efficiency of chlorpyrifos against chironomid midges and laboratory evaluation of substitute larvicides. *J. econ. Entomol.* **71**, 778–82.

Ali, A. & Mulla, M. S. (1978*b*). Effects of chironomid larvicides and diflubenzuron on non-target invertebrates in residential-recreational lakes. *Environ. Entomol.* **7**, 21–7.

Ali, A. & Mulla, M. S. (1979). Nuisance midge problem in Southern California, chemical control strategies. *Bull. Soc. Vector Ecol.* **4**, 44–53.

Anderson, D. R. & Burham, K. P. (1978). Effect of restrictive and liberal hunting regulations on annual survival rates of the mallard in North America. *Trans. N. Am. Wildl. Nat. Resour. Conf.* **43**, 181–6.

Anderson, D. W. & Keith, J. O. (1980). The human influence on seabird nesting success: conservation implications. *Biol. Conserv.* **18**, 65–80.

Arrington, N. & Edwards, A. E. (1951). Predator control as a factor in antelope management. *Trans. N. Am. Wildl. Nat. Resour. Conf.* **16**, 179–93.

Arthur, D. R. (1962). *Ticks and Disease.* Pergamon Press, Oxford.

Atkinson-Willes, G. L. (1969). Wildfowl and recreation: a balance of requirements. *British Water Supply*, **11**, 5–15.

Aung-Khin, M. (1980). The problem of snakebites in Burma. *Snake*, **12**, 125–7.

Baldridge, H. D. (1979). *Shark Attack.* Futura Publications, London.

Ball, D. F., Dale, J., Sheail, J. & Heal, O. W. (1982). *Vegetation Change in Upland Landscapes.* Institute of Terrestrial Ecology, Cambridge.

References

Barnes, E. S. (1973). Sewage pollution from tourist hotels in Jamaica. *Mar. Pollut. Bull.* **4** (7), 102–5.

Barr, S. E. (1974). Allergy to Hymenoptera stings. *J. Am. med. Ass.* **228**, 718–20.

Bates, G. H. (1935). The vegetation of footpaths, sidewalks, carttracks and gateways. *J. Ecol.* **23**, 470–89.

Batten, L. A. (1977). Sailing on reservoirs and its effects on water birds. *Biol. Conserv.* **11**, 49–58.

Baud-Bovy, M. & Lawson, F. (1977) *Tourism and Recreation Development – A Handbook of Physical Planning*. The Architectural Press Ltd, London.

Beale, C. J. & Wright, F. S. (1968). The Slimbridge observation hides. *Wildfowl.* **19**, 137–43.

Beazley, E. (1969). *Designed for Recreation*. Faber & Faber, London.

Belcher, H. & Swale, E. (1976). *A Beginner's Guide to Freshwater Algae*. Institute of Terrestial Ecology, H.M.S.O. London.

Bell, K. L. & Bliss, L. C. (1973). Alpine disturbance studies: Olympic National Park, USA. *Biol. Conserv.* **5**, 25–32.

Bell, R. H. V. (1971). A grazing system in the Serengeti. *Scient. Am.* **225**, 86–93.

Bellis, E. D. & Graves, H. B. (1971). Deer mortality on a Pennsylvania interstate highway. *J. Wildl. Mgmt* **35**, 232–7.

Bellrose, F. C. (1975). Impact of ingested lead pellets on waterfowl. *Proc. 1st Int. Wildfowl Symp.* 163–7. Ducks Unlimited, St Louis, Missouri.

Bertram, G. C. L. & Bertram, C. K. R. (1973). The modern Sirenia: their distribution and status. *Biol. J. Linn. Soc.* **5**, 297–338.

Billings, W. D. (1973). Arctic and alpine vegetations: similarities, differences and susceptibility to disturbance. *BioScience*, **23**, 697–704.

Blaškovič, D. (1967). The public health importance of tick-borne encephalitis in Europe. *Bull. Wld Hlth Org.* **36**, Suppl. 1, 5–13.

Boyd, H. & Lynch, J. L. (1984). Escape from mediocrity: a new approach to American waterfowl hunting regulations. *Wildfowl*, **35**, 5–13.

Breen, G. E. (1963). Unde venis? *Lancet*, i, 554.

British Birds (1982). Editorial. Codes for rarity-finders and twitchers. *Br. Birds*, **75**, 301–3.

British Herpetological Society (1984). *Toads on Roads Campaign*. Advisory leaflet.

British Medical Journal (1978a) Editorial: Ticks, tourists and encephalitis. *Br. med. J.* 587–8.

British Medical Journal (1978b). Editorial: Rocky Mountain spotted fever. *Br. med. J.* 651.

Brook, A. J. (1957). Water-blooms. *New Biol.* **23**, 86–101.

Brotherton, I., Maurice, O., Barrow, G. & Fishwick, A. (1977). *Tarn Hows, an Approach to the Management of a Popular Beauty Spot*. Countryside Commission, Cheltenham.

Brown, G. (1982). The Yellowstone perspective. *West. Wildlands*, 8, 28–30.

Brown, J. H. (1971). The desert pupfish. *Scient. Am.* **225**, 104–10.

Brown, J. H. & Lieberman, G. A. (1973). Resource utilisation and coexistence of seed-eating rodents in sand dune habitats. *Ecology*, **54**, 788–97.

Bruce-Chwatt, L. J. (1978). Mass travel and imported diseases. *Ann. Soc. belge Méd. trop.* **58**, 77–88.

Bruce-Chwatt, L. J. (1982). Imported malaria: an univited guest. *Br. med. Bull.* **38**, 179–85.

Budowski, G. (1976). Tourism and environmental conservation: conflict, coexistence, or symbiosis? *Environ. Conserv.* **3**, 27–31.

Burden, R. F. & Randerson, P. F. (1972). Quantitative studies of the effects of human trampling on vegetation as an aid to the management of semi-natural areas. *J. appl. Ecol.* **9**, 439–57.

Burdwick, G. E., Harris, E. J., Dean, T. M., Walker, J. S. & Colby, D. (1964). The accumulation of DDT in lake trout and the effect on reproduction. *Trans. Am. Fish. Soc.* **93**, 127–36.

Burgdorfer, W. (1975). A review of Rocky Mountain spotted fever (tick-borne typhus), its agent, and its tick vectors in the United States. *J. med. Ent.* **12**, 269–78.

Burkart, A. J. & Medlik, S. (1974). *Tourism – Past, Present, and Future.* Heinemann, London.

Bury, R. B., Luckenbach, R. A. & Busack, S. D. (1977). Effects of off-road vehicles on vertebrates in the California Desert. *U.S. Fish Wildl. Serv. Wildl. Res. Rep.* **8**, 1–23.

Cabelli, V. J. (1979). Evaluation of recreational water quality, the EPA approach. In *Biological Indicators of Water Quality* (ed. A. James & L. Evison), pp. 14/1–14/23. John Wiley & Sons, Chichester.

Cabelli, V. J. (1981). A health effects data base for the derivation of microbial guidelines for municipal sewage effluents. In *Coastal Discharges, Engineering Aspects and Experience* (Institution of Civil Engineers), pp. 51–5.

Cameron, A. M. (1981). Venomous fishes hazardous to humans. In *Animal Toxins and Man* (ed. J. Pearn), pp. 29–45. Division of Health Education and Information, Brisbane.

Canaway, P. M. (1980). Wear. In *Amenity Grassland: An Ecological Perspective* (ed. I. H. Rorison & R. Hunt), pp. 137–52. John Wiley & Sons, Chichester.

Cansdale, G. S. (1961). *West African Snakes.* Longman, London.

Carlander, K. D., Carlson, C. A., Gooch, V. & Wenke, T. L. (1967). Populations of *Hexagenia* mayfly naiads in Pool 19, Mississippi River, 1959–1963. *Ecology,* **48**, 873–8.

Carter, L. J. (1974). Off-road vehicles: a compromise plan for the California Desert. *Science, N.Y.* **183**, 396–9.

Čěrva, L. & Novák, K. (1968). Amoebic meningoencephalitis: sixteen fatalities. *Science,* **160**, 92.

Chan, K. L., Ho, B. C. & Chan, Y. C. (1971). *Aedes aegypti* (L.) and *Aedes albopictus* (Skuse) in Singapore City. 2. Larval habitats. *Bull. Wld Hlth Org.* **44**, 629–33.

Chappell, H. G., Ainsworth, J. F., Cameron, R. A. D. & Redfern, M. (1971). The effects of trampling on vegetation as an aid to the management of semi-natural areas. *J. appl. Ecol.* **8**, 869–82.

Chase, A. (1983). The last bears of Yellowstone. *Atlantic Monthly,* Feb. 1983, 63–73.

References

Cherfas, J. (1984). *Zoo 2000 – a Look Beyond the Bars*. British Broadcasting Corporation, London.

Chippindale, C. (1983). *The Complete Stonehenge*. Thames & Hudson, London.

Christiansen, M. I. (1977). *Park Planning Handbook*. John Wiley & Sons, New York.

Christie, A. B. (1980). *Infectious Disease: Epidemiology and Clinical Practice*. 3rd edn, Churchill Livingstone, Edinburgh.

Coblentz, B. E. (1978). The effects of feral goats (*Capra hircus*) on island ecosystems. *Biol. Conserv.* **13**, 279–86.

Coetzer, P. W. W. & Tilbury, C. R. (1982). The epidemiology of snakebite in northern Natal. *South Afr. med. Jour.* **62**, 206–12.

Cole, D. N. (1978). Estimating the susceptibility of wildland vegetation to trailside alteration. *J. appl. Ecol.* **15**, 281–6.

Connell, J. H. & Slatyer, R. O. (1977). Mechanisms of succession in natural communities and their role in community stability and organization. *Am. Nat.* **111**, 1119–44.

Connolly, G. E. (1980). Predators and predator control. In *Big Game of North America, Ecology and Management* (ed. J. L. Schmidt & D. L. Gilbert), pp. 369–94. Stackpole Books, Harrisburg, Pennsylvania.

Cook, S. F. (1965). The Clear Lake gnat, its control, past, present and future. *Calif. Vector Views*, **12**, 43–8.

Cook, S. F. (1981). The Clear Lake example: an ecological approach to pest management. *Environment*, **23**, 25–30.

Coppleson, V. M. (1950). A review of shark attacks in Australian waters since 1919. *Med. J. Aust.* **2**, 680–7.

Coppleson, V. M. (1962). *Shark Attack*. Halstead Press, Sydney.

Coppock, J. T. & Duffield, B. S. (1975). *Recreation in the Countryside – a Spatial Analysis*. Macmillan Press Ltd, London.

Cott, H. B. (1969). Tourists and crocodiles in Uganda. *Oryx*, **10**, 153–60.

Craighead, J. J. & Craighead, F. C. (1971). Grizzly bear–man relationships in Yellowstone National Park. *BioScience*, **21**, 845–57.

Davis, H. G., Eddy, G. W., McGovern, T. P. & Beroza, M. (1969). Heptyl butyrate, a new synthetic attractant for yellow jackets. *J. econ. Entomol.* **62**, 1245.

De Carneri, I. & Bianchi, L (1970). Infestazione collettiva da *Schistosoma haematobium* contratta da turisti italiani in Tunisia. *Gior. Mal. Infett. Parassit.* **22**, 640–5.

De Kadt, E. (ed.) (1979). *Tourism – Passport to Development?* Oxford University Press.

Department of Health and Social Security (1985). *Protect Your Health Abroad* Leaflet SA35. London.

De Silva, A. (1980). Snakebites and antivenom treatment in Sri Lanka. *Snake*, **12**, 134–7.

De Silva, A. (1981). Snakebites in Anuradhapura District. *Snake*, **13**, 117–30.

Dillon, P. J. & Rigler, F. H. (1975). A simple method for predicting the capacity of a lake for development based on trophic status. *J. Fish. Res. Bd Can.* **32**, 1519–31.

Domart, A., Gentilini, M. & Gaxotte, Ph. (1969). Fièvres de safari. A propos de 18 observations d'invasion bilharzienne. *Sem. Hôp. Paris*, **45**, 627–32.

Dorrance, M. J., Savage, P. J. & Huff, D. E. (1975). Effects of snowmobiles on white-tailed deer. *J. Wildl. Mgmt*, **39**, 563–9.

Dorst, J. (1970). *Before Nature Dies*. Collins, London.

Duffey, E., Morris, M. G., Sheail, J., Ward, L. K., Wells, D. A. & Wells, T. C. E. (1974). *Grassland Ecology and Wildlife Management*. Chapman & Hall, London.

Eckholm, E. P. (1975). The deterioration of mountain environments. *Science*, **189**, 764–70.

Edington, J. M. & Edington, M. A. (1977). *Ecology and Environmental Planning*. Chapman & Hall, London.

Edmonds, C. (1975). *Dangerous Marine Animals of the Indo-Pacific Region*. Wedneil Publication, Newport, Australia.

Ehrenfield, D. W. & Carr, A. (1967). The role of vision in the sea-finding orientation of the green turtle (*Chelonia mydas*). *Anim. Behav.* **15**, 25–36.

Endean, R. (1981). The box jelly-fish or "sea wasp". In *Animal Toxins and Man* (ed. J. Pearn), pp. 46–54. Division of Health Education and Information, Brisbane.

English Heritage (1985). *Stonehenge Study Group Report*. Historic Buildings and Monuments Commission for England.

Ennik, F. (1973). Abatement of yellowjackets using encapsulated formulations of Diazinon and Rabon. *J. econ. Entomol.* **66**, 1097–8.

Errington, P. L. (1946). Predation and vertebrate populations. *Q. Rev. Biol.* **21**, 144–77, 221–45.

Errington, P. L. (1967). *Of Predation and Life*. Iowa State University Press, Ames, Iowa.

Estes, J. A. & Palmisano, J. F. (1974). Sea otters: their role in structuring near shore communities. *Science*, **185**, 1058–60.

Estes, J. A., Smith N. & Palmisano, J. F. (1978). Sea otter predation and community organisation in the Western Aleutian Islands, Alaska. *Ecology*, **59**, 822–33.

Eyre, S. R. (1963). *Vegetation and Soils, a World Picture*. Edward Arnold, London.

Figley, H. D. (1940). Mayfly (Ephemeridae) hypersensitivity. *J. Allergy*, **11**, 376–87.

Forster, R. R. (1973). *Planning for Man and Nature in National Parks, Reconciling Perpetuation and Use*. International Union for Conservation of Nature and Natural Resources, Morges, Switzerland.

Frazier, C. A. (1969). *Insect Allergy: Allergic and Toxic Reactions to Insects and Other Arthropods*. Warren H. Green Inc., St Louis.

Fremling, C. R. (1968). Documentation of a mass emergence of *Hexagenia* mayflies from the upper Mississippi river. *Trans. Am. Fish. Soc.* **93**, 278–81.

Froment, A. (1981). Conservation of *Calluno-Vaccinietum* in the Belgian Ardennes, an experimental approach. *Vegetatio*, **47**, 193–200.

Fryer, G. & Iles, T. D. (1972). *The Cichlid Fishes of the Great Lakes of Africa: their Biology and Evolution*. Oliver & Boyd, Edinburgh.

Fuller, G. K., Lemma, A. & Haile, T. (1979). Schistosomiasis in Omo National Park of Southwest Ethiopia. *Am. J. trop. Med. Hyg.* **28**, 526–30.

References

Geddes, A. M. (1982). Intestinal infections in travellers. *Proc. R. Soc. Edinb.* (B), 82, 48–51.

Geis, A. D. (1963). Role of hunting regulations in migratory bird management. *Trans. N. Am. Wildl. Nat. Resour. Conf.* 28, 164–72.

Gilbert, P. W. (1963). The visual apparatus of sharks. In *Sharks and Survival* (ed. P. W. Gilbert), pp. 283–326. D. C. Heath & Co., Lexington.

Gildart, R. C. (1981). Roaming their last domains, grizzlies may clash with man. *Smithsonian*, 12, 78–87.

Giller, P. S. (1984). *Community Structure and the Niche*, Chapman & Hall, London.

Gilson, H. C. (1964). Lake Titicaca. *Verh. Internat. Verein. Limnol.* 15, 112–27.

Gimingham, C. H. (1972). *Ecology of Heathlands.* Chapman & Hall, London.

Godfrey, P. J. & Godfrey, M. M. (1980). Ecological effects of off-road vehicles on Cape Cod. *Oceanus*, 23(4), 56–67.

Golden, F. (1981). A plethora of polar bears. *Time Magazine*, Dec. 21, 1981, 46–7.

Gorham, P. R. (1964). Toxic algae. In *Algae and Man* (ed. D. F. Jackson), pp. 307–36. Plenum Press, New York.

Graber, D. & White, M. (1978). Management of black bears and humans in Yosemite National Park. *Cal-Neva Wildlife*, 1978, 42–51.

Grabherr, G. (1982). The impact of trampling by tourists on a high altitudinal grassland in the Tyrolean Alps, Austria. *Vegetatio*, 48, 209–17.

Grant, D. C., Rogers, C. J. & Lauret, T. H. (1968). Control of ground-nesting yellow jackets with toxic baits – a five-year testing program. *J. econ. Entomol.* 61, 1653–6.

Grassi, L. (1970). In margine all epidemia di schistosomiasi vesicale verificatasi tra turisti italiani in Tunisia. *Gior. Mal. Infett. Parassit.* 22, 646–7.

Great Barrier Reef Committee (1978). *Conservation and Use of the Capricorn Bunker Group of Islands and Coral Reefs.* Mimeographed Report, Great Barrier Reef Committee, Brisbane.

Great Barrier Reef Marine Park Authority (1980). *Great Barrier Reef Marine Park, Capricornia Section, Zoning Plan.* Townsville, Queensland.

Greenberg, B. (1965). Flies and disease. *Scient. Am.* 213(1), 92–9.

Griffin, J. L. (1972). Temperature tolerance of pathogenic and nonpathogenic free-living amoebas. *Science*, 178, 869–70.

Grodhaus, G. (1963). Chironomid midges as a nuisance, I. Review of biology. *Calif. Vector Views*, 10, 19–24.

Grodhaus, G. (1975). Bibliography of chironomid nuisance and control. *Calif. Vector Views*, 22, 71–80.

Grzimek, B. (1964). *Rhinos Belong to Everybody.* Collins, London.

Hall, F. (1972). Observations on blackflies of the genus *Simulium* in Los Angeles County, California. *Calif. Vector Views*, 19, 53–8.

Halliday, T. (1978). *Vanishing Birds, their History and Conservation.* Sidgwick & Jackson, London.

Halstead, B. W. (1978). *Poisonous and Venomous Marine Animals of the World.* The Darwin Press, Princeton, New Jersey.

Hansens, E. J. (1951). The stablefly and its effects on seashore recreational areas in New Jersey. *J. econ. Ent.* 44, 482–7.

Hansford, R. G. & Ladle, M. (1979). The medical importance and behaviour of *Simulium austeni* Edwards (Diptera: Simuliidae) in England. *Bull. ent. Res.* **69**, 33–41.

Hardman, J. A. & Cooper, D. R. (1980). Mute swans on the Warwickshire Avon – a study of a decline. *Wildfowl*, **31**, 29–36.

Harris, M. P. (1973). Evaluation of tourist impact and management in the Galapagos. In *World Wildlife Fund Yearbook* 1972–73, pp. 178–9. World Wildlife Fund, Morges, Switzerland.

Hartwick, R., Callanan, V. I. & Williamson, J. A. (1980). Disarming the box-jellyfish, nematocyst inhibition in *Chironex fleckeri*. *Med. J. Aust.* **1**, 15–20.

Harwood, R. F. & James, M. T. (1979). *Entomology in Human and Animal Health.* Macmillan Publishing Co., New York.

Hayashi, Y., Noburu, Y., Wada, Y. & Tanaka, H. (1979). The method to prevent the invasion of Habu (*Trimeresurus flavoviridis*) by an electric fence. *Snake*, **11**, 58–62.

Heinsohn, G. E. (1972). A study of dugongs (*Dugong dugong*) in Northern Queensland, Australia. *Biol. Conserv.* **4**, 205–13.

Herrero, S. (1976). Conflicts between man and grizzly bears in the National Parks of North America. In *Bears – Their Biology and Management* (ed. M. R. Pelton, J. W. Lentfer & G. E. Folk), pp. 121–45. International Union for Conservation of Nature and Natural Resources, Morges, Switzerland.

Holiday Which (1983). Illness abroad. *Holiday Which* May 1983, 147–51.

Hoogstraal, H. (1967). Tickborne hemorrhagic fevers, encephalitis, and typhus in U.S.S.R and Southern Asia. *Expl Parasit.* **21**, 98–111.

Hume, R. A. (1976). Reactions of goldeneyes to boating. *Br. Birds*, **69**, 178–9.

Hunt, E. G. & Bischoff, A. I. (1960). Inimical effects on wildlife of periodic DDD applications to Clear Lake. *Calif. Fish Game*, **46**, 91–106.

Hutchinson, M. (1985). Don't tame the mountains. *Countryside Commission News*, **13**, 3.

Institute of Terrestrial Ecology (1978). *Upland Land Use in England and Wales.* Countryside Commission Publication 111, Cheltenham.

IUCN (1968). Problems in species introductions. *Int. Union Conserv. Nature Nat. Resour. Bull.* **2**(9), 70–1.

IUCN (1976). *Proceedings of a Regional Meeting on the Creation of a Coordinated System of National Parks and Reserves in Eastern Africa.* Supplementary paper No. 45. International Union for Conservation of Nature and Natural Resources, Morges, Switzerland.

Jadin, J. B. (1981). Les aspects épidemiologiques et cliniques des amibes libres. *Ann. Station. Biol. Besse-en-Chandesse.* **15**, 1–14.

Janssens, P. G. & De Muynck, A. (1977). Malignant Rhodesian trypanosomiasis. *Ann. Soc. belge Méd. Trop.* **57**, 589–92.

Jarvinen, J. A. & Schmid, W. D. (1971). Snowmobile use and winter mortality of small mammals. In *Proceedings of 1971 Snowmobile and Off the Road Vehicle Research Symposium* (ed. M. Chubb), pp. 130–40. Michigan State University. Tech. Rep. No. 8. Recreation Research and Planning Unit, East Lansing.

References

Jenkins, D., Watson, A. & Miller, G. R. (1964). Predation and red grouse populations. *J. appl. Ecol* **1**, 183–95.

Jenkins, D., Watson, A. & Miller, G. R. (1967). Population fluctuations in the red grouse, *Lagopus lagopus scoticus. J. Anim. Ecol.* **36**, 97–122.

Johannes, R. E. (1971). How to kill a coral reef – II. *Mar. Pollut. Bull.* **2**(1), 9–10.

Johannes, R. E. (1975). Pollution and degradation of coral reef communities. In *Tropical Marine Pollution* (ed. E. J. Wood & R. E. Johannes), pp. 13–51. Elsevier Scientific Publishing Co., Amsterdam.

Johnson, R. H. & Nelson, D. R. (1973). Agonistic display in the gray reef shark, *Carcharhinus menisorrah*, and its relationship to attacks on man. *Copeia*, **1**, 76–84.

Jones, N. S. (1948). Observations and experiments on the biology of *Patella vulgata* at Port St Mary, Isle of Man. *Proc. Lpool. Biol. Soc.* **56**, 60–77.

Jungius, H. (1978). Plan to restore Arabian oryx in Oman. *Oryx*, **14**, 328–36.

Kadlec, V., Cěrva, L. & Škvárová, J. (1978). Virulent *Naegleria fowleri* in an indoor swimming pool. *Science*, **201**, 1025.

Kalmijn, A. J. (1971). The electric sense of sharks and rays. *J. exp. Biol.* **55**, 371–83.

Kalmijn, A. J. (1977). The electric and magnetic sense of sharks, skates and rays. *Oceanus*, **20**, 45–52.

Kar, S. K. & Bustard, H. R. (1983). Saltwater crocodile attacks on man. *Biol. Conserv.* **25**, 377–82.

Kaza, S. (1982). Recreational whalewatching in California: a profile. *Whalewatcher*, **16**(1), 6–8.

Kear, J. & Burton, P. J. K. (1971). The food and feeding apparatus of the Blue Duck *Hymenolaimus. Ibis*, **113**, 483–93.

Kettle, D. S., Reye, E. J. & Edwards, P. B. (1979). Distribution of *Culicoides molestus* (Skuse) (Diptera: Ceratopogonidae) in man-made canals in south-eastern Queensland. *Aust. J. mar. Freshwat. Res.* **30**, 653–60.

King, F. W. (1982). Historical review of the decline of the green turtle and the hawksbill. In *Biology and Conservation of Sea Turtles* (ed. K. A. Bjorndal), pp. 183–8. Smithsonian Institution Press, Washington, D.C.

Klauber, L. M. (1972). *Rattlesnakes, their Habits, Life Histories and Influence on Mankind*. University of California Press, Berkeley and Los Angeles.

Kohn, A. J., Saunders, P. R. & Wiener, S. (1960). Preliminary studies on the venom of the marine snail *Conus. Ann. N.Y. Acad. Sci.* **90**, 706–25.

Kruuk, H. (1972). Surplus killing by carnivores. *J. Zool. Lond.* **166**, 233–44.

Kury, C. R. & Gochfield, M. (1975). Human interference and gull predation in cormorant colonies. *Biol. Conserv.* **8**, 23–34.

La Bastille, A. (1974). *Ecology and Management of the Atitlán Grebe, Lake Atitlán Guatemala. Wildlife Monograph* **37**, Wildlife Society, Chestertown.

Lacey, L. A. & Federici, B. A. (1979). Pathogenesis and midgut histopathology of *Bacillus thuringiensis* in *Simulium vittatum* (Diptera: Simuliidae). *J. Invertebr. Pathol.* **33**, 171–82.

Laird, M. (ed), (1981). *Blackflies, the Future for Biological Methods in Integrated Control*. Academic Press, London.

Laming, A. (1959). *Lascaux, Paintings and Engravings*. Penguin Books, Harmondsworth.

La Scala, P. A. & Burger, J. F. (1981). A small-scale environmental approach to blackfly control in the U.S.A. In *Blackflies, the Future for Biological Methods in Integrated Control* (ed. M. Laird), pp. 133–6. Academic Press, London.

Laurence, B. R. & Mathias, P. L. (1972). The biology of *Leptoconops (Styloconops) spinosifrons* (Carter) (Diptera, Ceratopogonidae) in the Seychelles Islands, with descriptions of the immature stages. *J. med. Ent.* 9, 51–9.

Lawick-Goodall, H. van & Lawick-Goodall, J. van (1970). *Innocent Killers*. Collins, London.

Lee, R. B. (1972). The !Kung bushmen of Botswana. In *Hunters and Gatherers Today* (ed. M. G. Bicchieri), pp. 327–68. Holt, Rinehart & Winston, New York.

Lee, A. & Langer, R. (1983). Shark cartilage contains inhibitors of tumour angiogenesis. *Science*, 221, 1185–7.

Leeuwen, B. H. van (1982). Protection of migrating common toad (*Bufo bufo*) against car traffic in the Netherlands. *Environ. Conserv.* 9, 34.

Lefevre, M. & Laporte, G. S. (1969). The 'maladie verte' of Lascaux, diagnosis and treatment. *Stud. Speleol.* 2(1), 35–44.

Lent, P. C. (1974). Mother-infant relationships in ungulates. In *The Behaviour of Ungulates in Relation to Management* (ed. V. Geist & F. Walther), pp. 14–55. International Union for Conservation of Nature and Natural Resources, Morges, Switzerland.

Leppert, Z. (1982). *Ogwen*. Climbers' Club Guide to Wales.

Lewis, D. J. (1956). Chironomidae as a pest in the northern Sudan. *Acta trop.* 13, 142–58.

Liddle, M. J. (1975). A selective review of the ecological effects of human trampling on natural ecosystems. *Biol. Conserv.* 7, 17–36.

Liddle, M. J. & Scorgie, H. R. A. (1980). The effects of recreation on freshwater plants and animals: a review. *Biol. Conserv.* 17, 183–206.

Linley, J. R. & Davies, J. B. (1971). Sandflies and tourism in Florida and the Bahamas and Caribbean area. *J. econ. Ent.* 64, 264–78.

Lopez, B. H. (1978). *Of Wolves and Men*. J. M. Dent & Sons, London.

Lowday, J. E. & Wells, T. C. E. (1977). *The Management of Grassland and Heathland in Country Parks*. Countryside Commission, Cheltenham.

Luckenbach, R. A. (1975). What the ORVs are doing to the desert. *Fremontia*, 2(4), 3–11.

Luckenbach, R. A. (1978). An analysis of off-road vehicle use on desert avifaunas. *Trans. N. Am. Wildl. Nat. Resour. Conf.* 43, 157–62.

MacDonald, J. F., Akre, R. D. & Mathews, R. W. (1976). Evaluation of yellowjacket abatement in the United States. *Bull. ent. Soc. Am.* 22, 397–401.

Mack, D., Duplaix, N. & Wells, S. (1982). The sea turtle: an animal of divisible parts. International trade in sea turtle products. In *Biology and Conservation of Sea Turtles* (ed. K. A. Bjorndal), pp. 545–63. Smithsonian Institution Press, Washington D.C.

References

Mackay, F. (1982). Food, drink and travel. *Proc. R. Soc. Edinb.* (B), **82**, 37–45.

Mader, H. J. (1984). Animal habitat isolation by roads and agricultural fields. *Biol. Conserv.* **29**, 81–96.

Maegraith, B. (1963). Unde venis? *Lancet, i*, 401–4.

Mahoney, H. R. (1980). Dune busting: how much can our beaches bear. *Sea Frontiers*, **26**, 322–30.

Marine Mammal Commission (1982). *Report of a Meeting to Review On-going and Planned Research Concerning Humpback Whales in Glacier Bay and Surrounding Waters in Southeast Alaska.* U.S. Marine Mammal Commission, Washington D.C.

Mathieson, A. & Wall, G. (1982). *Tourism – Economic, Physical and Social Aspects.* Longman, London.

Mattingly, P. F. (1969). *The Biology of Mosquito-borne Disease.* Allen & Unwin, London.

McBride, J. R. (1974). Plant succession in the Berkeley Hills, California. *Madrono*, **22**, 317–29.

McBride, J. R. & Froehlich, D. (1984). Structure and condition of older stands in parks and open space areas of San Franciso, California. *Urban Ecology*, **8**, 165–78.

McBride, J. R. & Heady, H. F. (1968). Invasion of grassland by *Baccharis pilularis*, D. C. *Jour. Range Mgt*, **21**, 106–8.

McCloskey, M. (1983). *Recreational Whale-watching.* Mimeographed paper, The Whale Center, Oakland, California.

McFarlane, R. W. (1963). Disorientation of loggerhead hatchlings by artificial road lighting. *Copeia*, 1963, 153.

McLean, R. G. (1975). Racoon rabies. In *The Natural History of Rabies*, vol. 2 (ed. G. M. Baer), pp. 53–77. Academic Press, New York.

McMichael, D. F. (1971). Mollusks – classification, distribution, venom apparatus and venoms, symptomatology of stings. In *Venomous Animals and their Venoms*, vol. 3 (ed. W. Buecherl & E. E. Buckley), pp. 373–93. Academic Press, New York.

McQuaid-Cook, J. (1978). Effects of hikers and horses on mountain trails. *J. Environ. Manage.* **6**, 209–12.

Mech, L. D. (1970). *The Wolf: the Ecology and Behaviour of an Endangered Species.* Natural History Press, Garden City, New York.

Meggitt, M. J. (1964). Aboriginal food-gatherers of tropical Australia. In *The Ecology of Man in the Tropical Environment*, pp. 30–7. International Union for the Conservation of Nature and Natural Resources, Morges, Switzerland.

Merenne-Schoumaker, B. (1975). Aspects de l'influence des touristes sur les microclimats de la grotte de Remouchamps. *Ann. Speleol.* **30**, 273–85.

Meyer, J. (1981). Easy pickings. *Birds*, 8(6), 51–2.

Michalko, J. (1967). Vegetation on the southern slopes of the Tribec and Hronský Inovec Mountains. *Bull. Wld Hlth Org.* **36**, Suppl. I, 15–18.

Mikula, E. J., Martz, G. F. & Ryel, L. A. (1977). A comparison of lead and steel shot for waterfowl hunting. *Wildl. Soc. Bull.* **5**(1), 3–8.

Miller, R. R. (1950). Speciation in fishes of the genera *Cyprinodon* and *Empetrichthys* inhabiting the Death Valley region. *Evolution Lancaster, Pa.* **4**, 155–63.

Mills, J. A. (1976). Takahe and red deer. *Wildl. Rev. N.Z. Wildl. Serv.* **7**, 24–30.

Ministry of Housing and Local Government (1962). *Caravan Parks, Location, Layout, Landscape.* H.M.S.O., London.

Molloy, D. & Jamnback, H. (1981). Field evaluation of *Bacillus thuringiensis* var. *israelensis* as a black fly biocontrol agent and its effect on nontarget stream insects. *J. econ. Entomol.* **74**, 314–18.

Moment, G. B. (1968). Bears, the need for a new sanity in wildlife conservation. *BioScience*, **18**, 1105–8.

Moore, B. (1959). Sewage contamination of coastal bathing waters in England and Wales. A bacteriological and epidemiological study. *J. Hyg. Camb.* **57**, 435–72.

Moss, R., Watson, A. & Ollason, J. (1982). *Animal Population Dynamics.* Chapman & Hall, London.

Mount, G. A. (1981). Control of lone star tick in Oklahoma Parks through vegetative management. *J. econ. Entomol.* **74**, 173–5.

Mount, G. A. (1983). Area control of larvae of the lone star tick (Acari: Ixodidae) with acaricides. *J. econ. Entomol.* **76**, 113–16.

Mudge, G. P. (1983). The incidence and significance of ingested lead pellet poisoning in British wildfowl. *Biol. Conserv.* **27**, 333–72.

Mulla, M. S. (1974). Chironomids in residential-recreational lakes, an emerging nuisance problem – measures for control. *Ent. Tidskr.* **95**, (Suppl.) 172–6.

Munroe, E. G. (1951). Pest Trichoptera at Fort Erie, Ontario. *Can. Ent.* **83**, 69–72.

Murton, R. K. (1968). Some predator–prey relationships in bird damage and population control. In *The Problems of Birds as Pests* (ed. R. K. Murton & E. N. Wright). pp. 157–69. Academic Press, London.

Muscatine, L. & Cernichiari, E. (1969). Assimilation of photosynthetic products of zooxanthellae by a reef coral. *Biol. Bull.* **137**, 506–23.

Nature Conservancy Council (1979). *Wildlife Introductions to Great Britain.* Nature Conservancy Council, London.

Nature Conservancy Council (1981). *Lead Poisoning in Swans.* Nature Conservancy Council, London.

Newson, H. D. (1977). Arthropod problems in recreation areas. *Annu. Rev. Entomol.* **22**, 333–53.

Norris, K. S., Goodman, R. M., Villa-Ramirez, B. & Hobbs, L. (1977). Behaviour of California gray whale, *Eschrichtius robustus*, in Southern Baja California, Mexico. *Fishery Bull. Fish Wildl. Serv. U.S.* **75**, 159–72.

Nosek, J. & Grulich, I. (1967). The relationship between the tick-borne encephalitis virus and the ticks and mammals of the Tribeč mountain range. *Bull. Wld Hlth Org.* **36**, Suppl. 1, 31–47.

O'Connor, D. (1980). India's armoured giants. *Wildlife*, **22**(3), 40–2.

Odum, W. E. (1976). *Ecological Guidelines for Tropical Coastal Development.* International Union for Conservation of Nature and Natural Resources, Morges, Switzerland.

Ogden, J. C., Brown, R. A. & Salesky, N. (1973). Grazing by the echinoid *Diadema*

191

antillarum Philippi: formation of halos around West Indian patch reefs. *Science*, **182**, 715–17.

Owen, M. & Cadbury, C. J. (1975). The ecology and mortality of swans at the Ouse Washes, England. *Wildfowl*, **26**, 31–42.

Paine, R. T. (1966). Food web complexity and species diversity. *Am. Nat.* **100**, 65–75.

Paine, R. T. (1980). Food webs: linkage, interaction strength and community infrastructure. *J. Anim. Ecol.* **49**, 667–85.

Palmer, C. M. (1962). *Algae in Water Supplies*. U.S. Department of Health, Education and Welfare, Washington, D.C.

Parr, D. (1974). The effect on wildfowl of sailing at Island Barn Reservoir. *Surrey Bird Report* 1973, 74–8.

Parrish, H. M. (1959). Deaths from bites and stings of venomous animals and insects in the United States. *Archives Intern. Med.* **104**, 198–207.

Parrish, H. M., Goldner, J. C. & Silberg, M. A. (1965). Comparison between snakebites in children and adults. *Pediatrics*, **36**, 251–6.

Parrish, M. D. & Roberts, R. B. (1983). Insect growth regulators in baits: Methoprene acceptability to foragers and effect on larval eastern yellowjackets (Hymenoptera: Vespidae). *Ent. Soc. Am.* **76**, 109–12.

Parsons, D. J. (1983). Wilderness protection: an example from the southern Sierra Nevada, U.S.A. *Env. Conserv.* **10**, 23–30.

Paterson, R. (1979). Shark meshing takes a heavy toll of harmless marine animals. *Aust. Fish.* Oct. 1979, 17–23.

Patmore, J. A. (1983). *Recreation and Resources – Leisure Patterns and Leisure Places*. Basil Blackwell, Oxford.

Pavlovsky, E. N. (1966). *Natural Nidality of Transmissible Diseases with Special Reference to the Landscape Epidemiology of Zooanthroponeses*, transl. F. K. Plous & N. D. Levine. University of Illinois Press, Urbana.

Payne, R. & McVay, S. (1971). Songs of humpback whales. *Science*, **173**, 587–97.

Pearce, D. (1981). *Tourist Development*. Longman, London.

Pearce, P. L. (1982). *The Social Psychology of Tourist Behaviour*. Pergamon Press, Oxford.

Pelsue, F. W., McFarland, G. C. & Magy, H. I. (1970). Buffalo gnat (Simuliidae) control in the Southeast Mosquito Abatement District. *Proc. Calif. Mosquito Control Assoc.* **38**, 102–4.

Peters, W. (1982). Antimalarial drug resistance an increasing problem. *Br. med. Bull.* **38**, 187–92.

Peterson, D. G. (1952). Observations on the biology and control of pest Trichoptera at Fort Erie, Ontario. *Can. Ent.* **84**, 103–7.

Philibosian, R. (1976). Disorientation of hawksbill turtle hatchlings *Eretmochelys imbricata*, by stadium lights. *Copeia* 1976, 824.

Phillipson, J. (1966). *Ecological Energetics*. Edward Arnold, London.

Pienaar, U. de V. (1968). The ecological significance of roads in a National Park. *Koedoe*, **11**, 169–74.

Pigott, C. D. (1983). Regeneration of oak-birch woodland following exclusion of sheep. *J. Ecol.* **71**, 629–46.

Polunin, N. (1969). Conservation of the giant pied-billed grebe of Guatemala. *Biol. Conserv.* **1**, 176.

Pope, C. H. (1962). *The Giant Snakes.* Routledge & Kegan Paul, London.

Pratt, H. D. & Darsie, R. F. (1975). Highlights of medical entomology in 1974. *Bull. ent. Soc. Am.* **21**, 173–6.

Public Works (1961). Typhoid traced to bathing at a polluted beach. *Publ. Wks, N.Y.* **92**, 182–4.

Ranwell, D. S. (1972). *Ecology of Salt Marshes and Sand Dunes.* Chapman & Hall, London.

Rao, T. R., Trpis, M., Gillett, J. D., Teesdale, C. & Tonn, R. J. (1973). Breeding places and seasonal incidence of *Aedes aegypti*, as assessed by the single larva survey method. *Bull. Wld Hlth Org.* **48**, 615–22.

Ratcliffe, J. (1983). Why did the toad cross the road? *Wildlife*, **25**, 304–7.

Raven, P. H. (1966). *Native Shrubs of Southern California.* University of California Press, Berkeley.

Rees, S. H. (1974), *A Survey of Auk Colonies in North Wales.* Information Paper. Nature Conservancy Council, North Wales Region, Bangor.

Rehse-Küpper, B., Danielová, V., Klenk, W., Abar, B. & Ackerman, R. (1978). The isolation of Central European encephalitis (tick-borne encephalitis) virus from *Ixodes ricinus* (L.) ticks in Southern-Germany. *Zbl. Bakt. Hyg. I. Abt. Orig.* (A), **242**, 148–55.

Reid, B. (1969). Survival status of the Takahe, *Notornis mantelli*, of New Zealand. *Biol. Conserv.* **1**, 237–40.

Reid, D. (1982). Tourism and illness. *Proc. R. Soc. Edinb.* (B), **82**, 23–35.

Reid, H. A. (1983). Animal poisons. In *Manson's Tropical Diseases* (ed. P. E. C. Manson-Bahr & F. I. C. Apted), pp. 544–66.

Reid, H. A. & Theakston, R. D. G. (1983). The management of snake bite. *Bull. Wld Hlth Org.* **61**, 885–1028.

Ridgeway, R. (1982). Park at the top of the world. *Natn. geogr. Mag.* **161**, 704–25.

Rogers, J. P., Nichols, J. D., Martin, F. W., Kimball, C. F. & Pospahala, R. S. (1979). An examination of harvest and survival rates of ducks in relation to hunting. *Trans. N. Am. Wildl. Resour. Conf.* **44**, 114–26.

Ronsivalli, L. J. (1978). Sharks and their utilisation. *Mar. Fish. Rev.* **40**, 1–13.

Rosenberg, M. L., Hazlet, K. K., Schaefer, J., Wells, J. G. & Pruneda, R. C. (1976). Shigellosis from swimming. *J. Am. med. Ass.* **236**, 1849–52.

Ross Institute (1980). *Preservation of Personal Health in Warm Climates.* Ross Institute of Tropical Hygiene, London.

Ross Institute (1981). Malaria prevention in travellers from the United Kingdom. *Brit. med. J.* **283**, 214–18.

Russell, F. E. (1953). Stingray injuries: a review and discussion of their treatment. *Am. J. med. Sci.* **226**, 661–22.

Russell, F. E. (1965). Marine toxins and venomous and poisonous marine animals. *Adv. mar. Biol.* **3**, 255–384.

Saeed, A. A. & Magzoub, M. (1974). An outbreak of *Schistosoma mansoni* infection

193

in a European community in Darfur, Western Sudan. *Ann. trop. Med. Parasit.* **68**, 405–13.

Schaller, G. B. (1972). *The Serengeti Lion, a Study of Predator–Prey Relations.* University of Chicago Press, Chicago.

Schmidt, K. P. & Inger, R. F. (1957). *Living Reptiles of the World.* Garden City, New York.

Searle, W., Morel, G. J. & Hartwig, W. (1977). *A Field Guide to the Birds of West Africa.* Collins, London.

Sharp, A. (1977). *Gogarth.* Climbers' Club Guide to Wales.

Shaw, M. W. (1974). The reproductive characteristics of oak. In *The British Oak* (ed. M. G. Morris & F. H. Perring), pp. 162–81. E. W. Classey Ltd, Faringdon.

Sieveking, A. & Sieveking, G. (1962). *The Caves of France and Northern Spain: a Guide.* Vista Books, London.

Simenstad, C. A., Estes, J. A. & Kenyon, K. W. (1978). Aleuts, sea otters and alternate stable-state communities. *Science*, **200**, 403–11.

Simmons, I. G. (1975). *Rural Recreation in the Industrial World.* Edward Arnold, London.

Simpson, V. R., Hunt, A. E. & French, M. C. (1979). Chronic lead poisoning in a herd of Mute Swans. *Environ. Pollut.* **18**, 187–202.

Singer, P. (1975). *Animal Liberation, a New Ethics for Our Treatment of Animals.* Avon Books, New York.

Singh, D. (1967). The *Culex pipiens fatigans* problem in South-East Asia with special reference to urbanisation. *Bull. Wld Hlth Org.* **37**, 239–43.

Siow, K. T. & Moll, E. O. (1982). Status and conservation of estuarine and sea turtles in West Malaysian waters. In *Biology and Conservation of Sea Turtles* (ed. K. A. Bjorndal), pp. 339–47. Smithsonian Institution Press, Washington, D.C.

Smith, E. D. (1973). Electric anti-shark cable. *Civil. Eng. Pub. Wks. Rev.* Feb. 1973, 174–6.

Smith, E. D. (1974). Electro-physiology of the electrical shark-repellent. *Trans. S. Afr. Instn. elect. Engrs* **65**, 166–85.

Smith, E. D. (1979a). *A Report on Ecological Aspects of Electrical Anti-Shark Barriers.* National Physical Research Laboratory, Pretoria.

Smith, E. D. (1979b). *Cost Comparison Between Electrical Shark-Barriers and Meshing.* National Physical Research Laboratory, Pretoria.

Smith, M. (1982). How to save the forests of Snowdonia. *New Sci.* **95**, 14–17.

Smith, S. (1983). *Recreation Geography.* Longman, London.

Smith, V. L. (ed.) (1978). *Hosts and Guests, the Anthropology of Tourism.* Basil Blackwell, Oxford.

Soltz, D. L. & Naiman, R. J. (1978). The natural history of native fishes in the Death Valley system. *Nat. Hist. Mus. Los Ang. Cty Sci. Ser.* **30**, 1–76.

Southward, A. J. & Southward, E. C. (1978). Recolonisation of rocky shores in Cornwall after use of toxic dispersants to clean up the Torrey Canyon spill. *J. Fish. Res. Bd Can.* **35**, 682–706.

Spencer, C. (1975). Interband relations, leadership behaviour and the initiation of human-orientated behaviour in bands of semi-wild free ranging *Macaca fascicularis*. *Malay. Nat. J.* **29**, 83–9.

Spencer, C. (1979). The impact of man upon animal social behaviour: the example of altered social structure among macaque species. *Biol. Hum. Aff.* **44**, 7–14.

Spradberry, J. P. (1973). *Wasps, an Account of the Biology and Natural History of Solitary and Social Wasps.* Sidgwick & Jackson, London.

Stankey, G. H. (1978). Wilderness carrying capacity. In *Wilderness Management* (ed. J. C. Hendee, G. H. Stankey & R. C. Lucas), pp. 168–88. U.S. Dep. Agric. For. Serv., U.S. Government Printing Office, Washington, D.C.

Stark, N. (1969). Microecosystems in Lehman Cave, Nevada. *Natl. Speleol. Soc. Bull* **31**, 73–82.

Stebbins, R. C. (1974). Off-road vehicles and the fragile desert. *Am. Biol. Teach.* **36**, 203–8, 220–304.

Stevenson, A. H. (1953). Studies of bathing water quality and health *Am. J. publ. Hlth*, **43**, 529–38.

Stewart, D. R. M. (1963). The Arabian oryx (*Oryx leucoryx*) (Pallas). *E. Afr. Wildlife J.* **1**, 103–17.

Stidworthy, J. (1974). *Snakes of the World.* Grosset & Dunlap, New York.

Strand, S. (1972). Pack animal impact. In *Wilderness Impact Study Report* (ed. H. T. Harvey, R. J. Hartsveldt & J. T. Stanley), pp. 37–48. Sierra Club.

Strickland, D. (1981). The eskimo versus the walrus versus the government. *Nat. Hist.* **90**, (2), 48–57.

Swaroop, S. & Grab, B. (1954). Snakebite mortality in the world. *Bull. Wld Hlth Org.* **10**, 35–76.

Talbot, L. M. (1960). A look at threatened species. A report on some animals of the Middle East and Southern Asia which are threatened with extermination. *Oryx*, **5**, 153–293.

Tampion, J. (1977), *Dangerous Plants.* Universe Books, New York.

Taylor, V. (1977). Vanishing grey nurse shark. *Wildlife*, **19**, 450–3.

Tchernavin, V. V. (1944). A revision of the Subfamily Orestiinae. *Proc. zool. Soc. Lond.* **114**, 140–233.

Tester, A. L. (1963). Olfaction, gustation and the common chemical sense in sharks. In *Sharks and Survival* (ed. P. W. Gilbert), pp. 255–82. D. C. Heath & Co., Lexington.

Thomas, K. (1983). *Man and the Neutral World, Changing Attitudes in England 1500–1800.* Allen Lane, London.

Thompson, R. S., Burgdorfer, W., Russell, R. & Francis, B. J. (1969). Outbreak of tick-borne relapsing fever in Spokane County, Washington. *J. Am. med. Ass.* **210**, 1045–50.

Tuite, C. H. (1982). *The Impact of Water-based Recreation on the Waterfowl of Enclosed Inland Waters in Britain.* Sports Council and Nature Conservancy Council Report.

Van der Elst, R. P. (1979). A proliferation of small sharks in the shore-based Natal sport fishery. *Environ. Biol. Fishes*, **4**, 349–62.

References

Villwock, W. (1972). Gefahren für die endemische Fischfauna durch Einbürgerungs-versuche und Akklimatisation von Fremdfischen am Beispiel des Titicaca-Sees (Peru/Bolivien) und des Lanao-Sees (Mindanao/Philippinen). *Verh. Internat. Verein. Limnol.* 18, 1227–34.

Vollenweider, R. A. (1975). Input-output models with special reference to the phosphorus loading concept in limnology. *Schweiz. Z. Hydrol.* 37, 53–84.

Walker, E. & Williams, G. (1983). *ABC of Healthy Travel.* British Medical Association, London.

Wallace, J. B. & Merritt, R. W. (1980). Filter-feeding ecology of aquatic insects. *Ann. Rev. Entomol.* 25, 103–32.

Wallace, R. R. & Hynes, H. B. N. (1981). The effect of chemical treatments against blackfly larvae on the fauna of running waters. In *Blackflies, the Future for Biological Methods in Integrated Control* (ed. M. Laird), pp. 237–58. Academic Press, London.

Wallett, T. (1978). *Shark Attack and Treatment of Victims in Southern African Waters.* Purnell, Cape Town.

Walther, F. R. (1969). Flight behaviour and avoidance of predators in Thomson's gazelle (*Gazella thomsoni* Guenther 1884). *Behaviour*, 34, 184–221.

Warrell. D. A., Ormerod, L. D. & Davidson, N. McD. (1975). Bites by puff-adder (*Bitis arietans*) in Nigeria, and value of antivenom. *Br. med. J.* 4, 697–700.

Warrell, D. A., Ormerod, D. L. & Davidson, N. McD. (1976a). Bites by the night adder (*Causus maculatus*) and burrowing vipers (genus *Atractaspis*) in Nigeria. *Am. J. trop. Med. Hyg.* 25, 517–24.

Warrell, D. A., Greenwood, B. M., Davidson, N. McD., Ormerod, L. D. & Prentice, C. R. M (1976b). Necrosis, haemorrhage and complement depletion following bites by the spitting cobra (*Naja nigricollis*) *Q.Jl Med.* 45, 1–22.

Wayre, P. (1969). Wildlife in Taiwan. *Oryx*, 10, 46–56.

Weaver, T. & Dale, D. (1978). Trampling effects of hikers, motorcycles and horses in meadows and forests. *J. appl. Ecol.* 15, 451–7.

Webb, G. J. W., Yerbury, M. & Onions, V. (1978). A record of a *Crocodylus porosus* (Reptilia, Crocodylidae) attack. *J. Herpetol.* 12, 267–8.

Webb, R. H. (1982). Off-road motorcycle effects on a desert soil. *Environ. Cons.* 9, 197–208.

Wells, T. C. E. (1969). Botanical aspects of conservation of chalk grasslands. *Biol. Conserv.* 2, 36–44.

Wells, T. C. E. (1970). A comparison of the effects of sheep grazing and mechanical cutting on the structure and botanical composition of chalk grassland. In *The Scientific Management of Animal and Plant Communities for Conservation* (ed. E. Duffey & A. S. Watt), pp. 497–515. Blackwell Scientific Publications, Oxford.

Wells, T. C. E. (1980). Management options for lowland grassland. In *Amenity Grassland: An Ecological Perspective* (ed. I. H. Rorison & R. Hunt), pp. 175–95. John Wiley & Sons, Chichester.

Went, F. W. (1955). The ecology of desert plants. *Scient. Am.* 192, 68–75.

Willaert, E. (1974). Primary amoebic meningo–encephalitis. A selected bibliography and tabular survey of cases. *Ann. Soc. belge Med. trop.* 54, 429–40.

Willard, B. E. & Marr, J. W. (1970). Effects of human activities on alpine tundra ecosystems in Rocky Mountain National Park, Colorado. *Biol. Conserv.* **2**, 257–65.

Willard, B. E. & Marr, J. W. (1971). Recovery of alpine tundra under protection after damage by human activities in the Rocky Mountains of Colorado. *Biol. Conserv.* **3**, 181–90.

Willems, J. H. (1983). Species composition and above ground phytomass in chalk grassland with different management. *Vegetatio*, **52**, 171–80.

Williamson, J. A., Callanan, V. I. & Hartwick, R. F. (1980). Serious envenomation by the northern Australian box-jellyfish (*Chironex fleckeri*). *Med. J. Aust.* **1**, 13–15.

Wilshire, H. G., Shipley, S. & Nakata, J. K. (1978). Impacts of off-road vehicles on vegetation. *Trans. N. Am. Wildl. Nat. Resour. Conf.* **43**, 131–9.

Wodzicki, K. A. (1961). Ecology and management of introduced ungulates in New Zealand. *Terre Vie*, **1**, 130–57.

Woodruff, A. W. (1975). Diseases of travel, with particular reference to tropical diseases. *Post-grad. med. J.* **51**, 825–9.

World Health Organisation (1972). *Vector Control in International Health.* Geneva.

Zaret, T. M. & Paine, R. T. (1973). Species introduction in a tropical lake. *Science*, **182**, 449–55.

Zaslowsky, D. (1983). "Black cavalry of commerce", hotels, hot dogs, and the concessioner syndrome. *Wilderness*, **46**, (160), 25–32.

Zdarek, J., Haragsim, O. & Vesely, V. (1976). Action of juvenoids on the honey bee colony. *Z. angew. Ent.* **81**, 392–401.

Zuidema, P. J. (1981). The Katayama syndrome; an outbreak in Dutch tourists to Omo National Park, Ethiopia. *Trop. geogr. Med.* **33**, 30–5.

Zumpt, F. (1973), *The Stomoxyine Biting Flies of the World (Diptera: Muscidae). Taxonomy, Biology, Economic Importance and Control Measures.* Gustav-Fischer Verlag, Stuttgart.

INDEX

Acaricides, 110
Active physical pursuits, 13–33
Airports
    malaria hazard at, 103–4
Algae
    as cause of: bathing problems, 12, 173; cave disfigurement,
        84–6; coral reef destruction, 175–6; fish mortality, 173
    relationship with sewage effluents, 12, 173–7   (see also
        seaweeds)
Allergic reactions
    to insect bites, 126
    to insect remains, 118, 120, 123
    to wasp stings, 131
Alpine vegetation
    damage by: climbers, 20–3; horse riders, 22–3; pack animals,
        23–5; walkers, 78–80
Angling, 71–4
Animal introductions, 61–5
Animal watching, 34–50
Arabian oryx, 53–4
Archaeological sites, 82–6
Artificial feeding of animals
    as attraction to viewing areas, 48
    as cause of aggressive behaviour, 35–7, 161–5
    as cause of territorial disruption, 41

Baboons, 36–7, 38–40
Bacillus thuringiensis, 127
Bacteriological standards
    for bathing waters, 113–15
Badger, 169, 171
Bathing
    hazards from: jellyfish, 147–9; sharks, 135–46; stingrays,
        146–7
    health risks associated with sewage pollution, 111–15
    water standards, 113–15
Bears
    attacks on humans, 163
    feeding on refuse, 162, 177
    habituation to human foodstuffs, 162–5
Bird watching
    disturbance caused by, 34–5
    hides used for, 46
Biting flies
    control of, 126–30
    types of: blackflies, 125–7; clegs, 124; deer flies, 124;
        horseflies, 124; mosquitoes, 100–4; sandflies (biting
        midges), 127–30; stable flies, 130–1, 173
Blackflies (Simuliidae), 125–7
Blanket weed (Cladophora), 173–4
Bonding period, 42–3
Brown pelican, 38–9

Caddis flies, 118–19, 122–4
Calcite, 84–5
Camping, 23–5, 31, 105, 109
Carrying capacity, 80
Cattle, 90

Caves, 84–5
Chaoborid midges, 121
Cheetah, 40
Chironomid midges, 118–22
Cities in tropics
    health hazards, 100–3
Clear Lake gnat, 121
Climbing
    and seabird disturbance, 15–17
    and vegetation damage, 20–3
    restrictions on, 17
Coliform counts, 113–15
Communication problems
    in relation to health hazards, 116–17
Community structure
    disturbances of, 7–12, 64–8, 152
Competition
    definition, 5–6
    introduced and native species, 61–8
Coral reefs
    biological hazards of, 150–3
    effect of sewage on, 175–7
Country Parks, 89–91
Coyote, 58
Coyote brush (Baccharis), 87–9
Crocodiles
    attacks on man, 153–4
    disturbance by tourists, 38–40

Deer
    and forest damage, 67, 75
    as disease reservoirs, 105–6
    mortality on roads, 169
    reaction to snowmobiles, 13
    use in vegetation management, 89–91
Desalination plants, 180
Deserts
    damage from off-road vehicles, 26–31
    flora and fauna of, 26–30
    zoning plan for, 30–1
Disease hazards
    amoebiasis, 96
    bacillary dysentry, 112, 115
    dengue fever, 101–2, 178
    filariasis, 100
    gastroenteritis, 112–15
    leptospirosis, 177
    malaria, 100, 102–4
    meningo-encephalitis, 115–16
    paratyphoid fever, 113–15
    rabies, 36
    schistosomiasis, 97–100
    sylvatic plague, 36
    tick-borne, 105–10
    travellers diarrhoea, 96
    trypanosomiasis, 96
    typhoid fever, 112, 114–15
    viral hepatitis, 113–15

yellow fever, 6, 102, 178
Disease reservoirs
  definition of, 5–6
  examples of: chipmunks, 107, 109; deer, 105–6; goats, 106;
    hedgehogs, 111; monkeys, 5; rats, 177; squirrels, 36,
    107, 109
Disease vectors
  definition of, 5–6
  mosquito species: *Aedes aegypti*, 101–2; 178; *Aedes albopictus*,
    101–2; *Anopheles gambiae*, 100–1; *Anopheles stephensi*,
    100; *Culex pipiens fatigans*, 100–2
  tick species: *Amblyomma americanum*, 108–9; *Dermacentor*
    andersoni, 107–8; *Dermacentor variabilis*, 108–9; *Ixodes*
    ricinus, 105–6; *Ornithodoros hermsi*, 109–10
Ducks
  and reservoir use, 18–21
  lead poisoning in, 71–2, 74
Duck shooting, 55–6
Dugong, 141

East African safaris
  artificial feeding problems, 37
  snake hazards, 156–7
  viewing lodges, 47–8
  wildlife disturbance, 39–40, 42–3
Ecological relationships
  classification of, 5–12
Ecology
  definition of, 4
Ecosystem
  definition of, 7
Elephants
  artificial feeding of, 37
  use for animal viewing, 47
Energy flow in communities, 11–12
Energy pyramids, 11–12
Enjoyment of scenery, 77–95
Eutrophication, 123
Evolution
  communities demonstrating, 61–5

Firewood
  use by tourists, 179–80
Fish
  introduced species: brown trout, 65, 67; largemouth bass, 61,
    64, 66; peacock bass, 8–9; rainbow trout, 64
  faunas of: African lakes, 65; coral reefs, 150–2; Death Valley,
    61–4; Gatun Lake, 8–9; kelp beds, 9–10; Lake Titicaca,
    65
Fishing
  and discarded lines, 71
  and exotic species, 61–7
  and lead weights, 71–5
  and predator control, 58–9
  and toxic algae, 173
Flyways
  used by migrating waterfowl, 55–6
Food chains, 5–12, 121–2
Food storage
  and bears, 164–5
Footpath wear, 78–84
Foxes, 29, 58, 75

Galápagos Islands, 8, 41
Game birds
  bobwhite quail, 57
  ducks, 54–6
  pheasants, 58
  red grouse, 57
Game management, 54–60
Game animal introductions, 61–8
Gastrointestinal disorders, 96, 114, 173
Gazelles, 6, 41
Grassland management, 80, 87–92
Gray whale, 43–5
Grazing
  and plant succession, 86–7
  and tree regeneration, 93–4
  and vegetation damage, 23–5, 68
  and vegetation management, 87–93
  by marine animals, 8–10, 152
Great Barrier Reef, 45, 178–9

Grebes, 65–6, 121–2
Green bubble alga (*Dictyosphaeria*), 174–6
Growth hormones
  use in insect control, 122, 133–4
Gulls
  association with refuse, 177–9
  predation on other seabirds, 15, 37–9, 179

Heron, 58–9
Horseflies, 124
Horse riding
  and vegetation damage, 22–5
Horses, 22–5, 90
Houseflies, 131, 178
Huia, 68–9
Hunting and shooting
  associated with souvenir trade, 68–71
  opposition to, 75–6
  pest control value of, 75
  predator control associated with, 56–60
  regulations governing, 55–6
  species diminished by, 52–4

Immunisation, 102, 111
Indicator organisms and sewage pollution, 114–15
Insect
  attractants, 132–3
  control by: fish, 122; growth hormones, 112, 133–4; habitat
    modification, 129–30; insecticides, 102, 121, 124, 129,
    131, 133–4
  repellents, 129
Insecticides
  development of resistance to, 102, 122
  organochlorine, 121, 126
  organophosphorus, 122, 126
  side effects on honeybees, 134

Jellyfish, 147–9

Kakapo (flightless parrot), 67–8
Kangaroo rat, 29
Keystone species, 8–11, 142, 152

Lakes
  as waterbird refuges, 18–21, 54–6
  midge problems associated with 120–2
  nutrient enrichment of, 12, 123, 176
  recreational use of, 18–21, 64–6
  zoning of, 19–21
Lake Atitlán, 65–6
Lake Gatun, 8–9
Lake Merritt, 21
Lake Titicaca, 65
Lascaux, 84–6
Lead-poisoning in waterbirds, 71–5
Lights
  and turtles, 170–1
  in caves, 84–5
Lion, 6, 40, 48
Lizards
  desert, 28–9
  land iguana, 41
  monitor, 38, 40
Log cabins
  and tick-borne diseases, 109–10

Macaque, 36–7
Malaria
  drug resistance of parasite, 103
  mosquito vectors of, 100–1
  occurrence amongst tourists, 96, 103–4
  prophylactic measures against, 102–3
  risk at airports, 103–4
Manatee, 14
Mangrove swamps, 130
Marabou storks, 177–8
Mayflies, 118–19, 122–3
Midges, 118–22, 127
Minibuses
  and wildlife disturbance, 43, 49
Monkeys, 5–6, 37
Moorland management, 92

Index

Mosquitoes, 100–4
 (*see also* disease vectors)
Moss, 84
Motor boats as cause of
 damage to aquatic plants, 13
 injury to manatees, 13–14

Nutrient enrichment of lakes, 12, 123, 176

Observing wildlife, 34–50
Off-road vehicles, 13–14, 26–32

Parent/offspring bonds in animals
 disruption by tourists, 42–5
Pheasants, 58, 69–70
Picnic sites
 problems from wasps, 131
 vegetation management, 89
Plant succession, 86–93
Poison oak, 87–9
Population dynamics, 7, 56–7, 141–2
Predator/prey relationships
 definition of, 5–7
 disturbance by tourists, 37–40, 179
Predators
 control by recreational hunters, 56–60
Pronghorn antelope, 52, 58

Racoon, 36
Rats
 and leptospirosis transmission, 177
 and refuse, 177–9
 as attraction for snakes, 157
Refuse disposal, 162, 177–9
Reservoirs
 interaction between sailing and waterbirds, 18–21
Restrictions designed to limit recreational disturbance
 camping bans, 23–5
 definition of safe approach distances, 49–50
 entry permits, 25
 seasonal closures, 17–20, 45
 zoning, 19–21, 30–1, 45–6
Roads
 animal mortality on, 169–70
 as barrier to animals, 167–9
Road underpasses for animal use, 171–2
Rock type
 and plant distribution, 21–3

Sailing
 and waterbirds, 18–21
Saltlicks, 46–7
Sand dunes (coastal)
 and coastal protection, 31
 and water supply, 180
 damage by off-road vehicles, 26–32
Sandflies, 127–30
Schistosomiasis (Bilharzia), 97–100
Seabirds
 disturbance by climbers, 15–17
 disturbance by tourists, 37–9
Sea otter, 8–11
Sea-urchins
 as algal grazers, 8–10, 152
 as hazard on coral reefs, 151–2
Seaweed
 grazing by limpets, 8
 grazing by sea-urchins, 8–10, 152
Service facilities, 77, 167–80
Sewage pollution
 and algal growths, 12, 173–7
 and health hazards, 111–15
Sewage treatment, 176–7
Sharks
 attacks on bathers, 135–46
 barriers against, 143–5
 benefits to man, 145–6
 control by netting, 140–3
Sheep, 89–93
Snakebite, 154–60
Snakes, 29, 153, 155–60
Snowmobiles

 disturbance caused by, 13–14
Soil disturbance
 resulting from recreation, 13, 26–8, 78, 81–2
Souvenir trade, 68–71
Spear-fishing and shark attacks, 136–7
Squirrels, 36, 109–10
Stable flies, 130–1, 173
Stingrays, 146–7
Stock grazing
 and grassland management, 87–93
 and suppression of tree regeneration, 93–4
Stonehenge, 82–4
Support facilities, 167–80
Swans, 72–4
Swimming pools, 99, 115–16

Takahe (flightless rail), 67–8
Territorial behaviour
 disrupted by tourists, 41
 implicated in shark attacks, 136
Ticks
 as disease vectors, 104–10
 control measures against, 110–11
Toads
 mortality on roads, 170, 172
Tortoises, 8, 27–9, 167
Trees
 as scenic features, 94–5
 as screens, 77
 damage to roots by trampling, 81–2
 suppression of regeneration by: competitive ground
  vegetation, 95; shading, 94; stock grazing, 93–4
Trail bikes, 26–31
 (*see also* off-road vehicles)
Trampling damage, 22–5, 78–84
 (*see also* vegetation damage)
Turtles
 as tourist souvenirs, 70–1
 disoriented by lights, 170–1
 trapped in shark nets, 141

Vaccines, 102, 111
Vegetation damage
 caused by: climbing, 20–3; horse riding, 22–5; motor boats,
  13; off-road vehicles, 26–32; pack animal grazing, 22–5;
  walking, 14, 78–84
 remedial measures: artificial surfacing, 81–2; fertilising and
  reseeding, 80; realigning of footpaths, 81–2
Venomous animals
 protective measures against, 149, 152–3, 159–60
 types of: cone shells, 150–1; coral reef fish, 150–2; jellyfish,
  147–9; sea-urchins, 151–2; snakes, 154–60; stingrays,
  146–7; wasps (yellow jackets), 131–4
Viewing facilities for wildlife, 46–8

Walrus, 70
Warm-water amoebae, 115–16
Wasps, 131–4
Water supply, 180
Waterbirds
 affected by insecticides, 121–2
 affected by lead-poisoning, 71–5
 disturbed by sailing, 18–21
 entangled in fishing lines, 71
 zoning to protect, 19–20
Water-blooms, 173
Waterholes, 46–8
Water table, 129–30, 180
Whale watching, 43–5, 49–50
Wildebeest, 6, 42–3
Wildlife souvenir trade, 68–71
Wolves, 58, 153
Woodland, 93–5

Yellowjackets, 131–4

Zebra, 6, 43
Zoning to minimise recreational disturbance
 in Bialowieza National Park, 45–6
 in California desert, 30–1
 on Great Barrier Reef, 45
 on lakes and reservoirs, 19–21